THE STEEL MARKET
IN 1996
AND THE OUTLOOK
FOR 1997 AND 1998

ORGANISATION FOR ECONOMIC CO-OPERATION AND DEVELOPMENT

ORGANISATION FOR ECONOMIC CO-OPERATION AND DEVELOPMENT

Pursuant to Article 1 of the Convention signed in Paris on 14th December 1960, and which came into force on 30th September 1961, the Organisation for Economic Co-operation and Development (OECD) shall promote policies designed:

- to achieve the highest sustainable economic growth and employment and a rising standard of living in Member countries, while maintaining financial stability, and thus to contribute to the development of the world economy;
- to contribute to sound economic expansion in Member as well as non-member countries in the process of economic development; and
- to contribute to the expansion of world trade on a multilateral, non-discriminatory basis in accordance with international obligations.

The original Member countries of the OECD are Austria, Belgium, Canada, Denmark, France, Germany, Greece, Iceland, Ireland, Italy, Luxembourg, the Netherlands, Norway, Portugal, Spain, Sweden, Switzerland, Turkey, the United Kingdom and the United States. The following countries became Members subsequently through accession at the dates indicated hereafter: Japan (28th April 1964), Finland (28th January 1969), Australia (7th June 1971), New Zealand (29th May 1973), Mexico (18th May 1994), the Czech Republic (21st December 1995), Hungary (7th May 1996), Poland (22nd November 1996) and the Republic of Korea (12th December 1996). The Commission of the European Communities takes part in the work of the OECD (Article 13 of the OECD Convention).

Publié en français sous le titre :

LE MARCHÉ DE L'ACIER EN 1996 ET LES PERSPECTIVES POUR 1997 ET 1998

FOREWORD

The OECD Steel Committee decided to undertake studies of the stell market and its outlook at its first meeting in 1978. Since that date, a report has been published yearly, beginning with *The Steel Market in 1978 and the Outlook for 1979 and 1980.*

This report was prepared by Mr. Franco Mannato of the OECD Secretariat. The Steel Commitee examined the report, which is based on data received before 31 March 1997. It is published on the responsibility of the Secretary-General of the OECD.

TABLE OF CONTENTS

INTRODUCTION

At its 48th meeting in May 1996 the Steel Committee decided, when adopting its programme of work, that a report on steel market trends in 1996 and the outlook for 1997 and 1998 would be drawn up in early 1997. After being discussed by the Steel Committee, the report will be published under the responsibility of the Secretary-General, as in previous years.

Certain delegations to the Steel Committee provided statistics and other information on market developments in their countries, and the Secretariat has taken these into account. Since, however, it had to produce a coherent world outlook, it is possible that the text and the estimates may differ somewhat from those provided by the various delegations. It is therefore the Secretariat which is responsible for the present document.

The present report is broadly along the same lines as its predecessors, the most recent being the 1996 report entitled: *The Steel Market in 1995 and the Outlook for 1996 and 1997.*

Following the accession of the Republic of Korea, Hungary, Poland and the Czech Republic as Member countries of the OECD during 1996, the statistics for these countries have been included in the OECD total and, as far as possible, in order to maintain a degree of coherence, the historical data have been recalculated on that basis. Also, since Brazil became a full participant in the Steel Committee in 1996, statistics for that country have been added to most of the tables and removed from those for the Latin America zone.

As a result of these changes, the data for Hungary, Poland and the Czech Republic have been removed from the central and eastern European zone and have been included in the "Other Europe" zone. As regards the European Union, only the EU (15) zone remains and the historical data have been recalculated as far as possible.

The report has been drawn up using the information received and the statistics available as at 31 March 1997.

NOTES ON THE MAIN FEATURES OF THE STEEL MARKET IN 1996, 1997 AND 1998

The main quantitative results for the steel market in 1996, and probable trends in 1997 and 1998, are contained in the statistical annex. The main developments in this market may be summarised as follows:

1996

APPARENT STEEL CONSUMPTION

- World: world steel consumption, which had been rising since 1993, levelled out in 1996, declining by 0.2 per cent compared to 1995. This situation reflects the fall in consumption in the OECD area and that which continued in the New Independent States (ex-USSR); these declines were compensated by the rises in the non-OECD market-economy countries, particularly China.

- OECD: following the distinct upturn in apparent steel consumption in the OECD area in 1994 (+ 9.4 per cent) which continued in 1995 (+6.7 per cent), demand for steel fell by 3.6 per cent in 1996 in the OECD area as a whole, representing a drop of 14.4 million tonnes in finished product equivalent. This substantial fall in apparent consumption seems to have been largely due to the considerable stock drawdowns at the end of 1995 which continued in the first half of 1996.

- Among Member countries, the United States, Mexico and Korea recorded increases in apparent steel consumption, of 6.2, 6.3 and 3.7 per cent respectively. In the other OECD countries, particularly sharp falls were recorded in Europe, a decline of more than 12 per cent in EU (15) and over 16 per cent for the other European countries. In Oceania and Canada the falls were much less substantial and in Japan the reduction in apparent steel consumption was only 1.2 per cent.

- In Brazil, a participant in the OECD Steel Committee, steel consumption – after increasing over a four-year period – levelled out, and fell by 0.6 per cent in 1996 compared with 1995.

- The demand for steel in the other market-economy areas continued to rise for the third year running, but at a slower rate than in previous years. The increase was of the order of 4.8 per cent, equivalent to 5.2 million tonnes. Excluding Africa (but not South Africa), where steel consumption again fell by 5.2 per cent, steel consumption rose in every other region.

- Contrary to expectations, the downward trend in the New Independent States (ex-USSR) did not slow down. The fall continued at a high rate of 9.6 per cent and steel consumption will have been about 30 million tonnes, equivalent to hardly more than a quarter of apparent steel consumption in the USSR in 1988.

- In the central and eastern European countries which did not become members of the OECD, steel consumption continued to rise in 1996, by 5.6 per cent over 1995. Rises of nearly 56 per cent in the Slovak Republic and of nearly 12 per cent in Romania largely compensated for a fall of nearly 50 per cent in Bulgaria.

- In China and North Korea, following the drop in 1995 (–6.4 per cent), demand for steel went up by 11.2 per cent, or nearly 11 million tonnes more than in 1995.

Growth in apparent consumption of steel (AC) and estimated growth in real consumption of steel (RC) and total steel stocks held by steel producers consumers and merchants [1]

	OECD AC	Rest of OECD AC	Total for the United States, the European Union and Japan				
			AC	RC	Total stocks of steel		
					Yearly change	End-year level	
			In million tonnes of finished product equivalent				In weeks of real consumption
1985	282.7	41.2	241.5	246.3	−4.8	78.2	16.5
1986	274.8	41.1	233.7	239.6	−5.9	72.3	15.7
1987	289.0	43.1	245.9	243.9	+2.0	74.3	15.9
1988	324.6	45.9	278.7	261.1	+17.6	91.9	18.3
1989	330.4	46.2	284.2	279.9	−4.3	96.2	17.9
1990	330.8	40.3	290.5	286.8	+3.7	99.9	18.1
1991	316.7	38.5	278.2	281.9	−3.7	96.2	17.7
1992	310.9	40.7	270.2	272.3	−2.1	94.1	18.0
1993	306.0	44.4	261.6	264.2	−2.6	91.5	18.0
1994	334.2	49.5	284.7	284.7	0	91.5	16.7
1995	396.3	92.2	304.1	298.4	+5.6	97.1	16.9
1996	381.9	88.7	293.2	299.4	−6.2	90.9	15.8
1997e	388.5	92.3	296.1	296.8	−0.6	90.3	15.8
1998p	403.1	97.3	305.8	305.3	+0.6	90.9	15.5

e: Estimate.
p: Forecast.
1. In previous years, figures for apparent steel consumption can be derived, as they have in this report, from the data available on steel production and trade. Variations in apparent consumption are due to variations in real consumption and/or changes in the total steel inventories maintained by steel producers, consumers and merchants. Data regarding the level of, or annual variations in, both these parameters, however, are far from complete. The figures given for real consumption and annual variations in total stock levels should therefore be taken as "reasonable" estimates of two interrelated factors. Furthermore, in calculating the level of total steel stocks in tonnage terms by the end of 1984, it has been assumed that the stocks were equal to 18 weeks of estimated real consumption for that year (i.e. 8 weeks for producers and 10 weeks for consumers and merchants). For the years after 1984, the level of total steel stocks was first calculated in terms of tonnage, based on the estimated annual variation, and subsequently expressed in terms of weeks of real consumption.
2. As from 1995, data concerning the EU are related to EU (15) and the OECD total includes the new Member countries: Korea, Hungary, Poland and the Czech Republic.
Source: OECD Secretariat.

– Overall, steel stocks in the OECD area fell substantially during 1996, this phenomenon being particularly marked in Europe, particularly in the European Union, while stocks went up in the United States. As a result of the trend in steel stocks during 1996, real steel consumption in the OECD area as a whole should have risen by 0.5 per cent over the year.

STEEL TRADE

– World trade in steel (excluding intra-EU trade) appears to have steadied out at nearly 165 million tonnes, only 0.7 per cent down on 1995. World trade in steel accounted for nearly 25.5 per cent of world steel consumption.

– Steel exports from the OECD area (as it stood in 1996) increased by 0.9 per cent in 1996, totalling 0.8 million tonnes more than in 1995. Imports fell by more than 6 million tonnes, or 7.8 per cent. Consequently net OECD exports went up by 50 per cent.

– Steel imports in the United States, after falling in the first half of 1996, rose substantially in the second half of the year to a total of 26.5 million tonnes for the year, or 19.5 per cent up on 1995. As a result of problems encountered in a number of plants, imports of semi-finished products went up by 55 per cent. However exports of steel products fell by 29 per cent from the record level they reached in 1995, despite a healthy domestic market, and the share of imports on the American market went up considerably from 22.6 per cent in 1995 to 25.9 per cent in 1996.

- Net exports from the European Union appear to have risen considerably, by more than 60 per cent in 1996 compared with 1995, being the combined effect of a 35 per cent fall in imports and a 4.3 per cent rise in exports.

- Japanese imports of steel fell by 15.4 per cent in 1996 to a more normal level, following the marked increase in 1995. At the same time, exports fell by 12.3 per cent, causing Japanese net exports to fall by 10.8 per cent. The share of imports on the Japanese market fell from 8.8 per cent in 1995 to 7.5 per cent in 1996.

- Net steel imports by non-OECD market economies increased in 1996 by a further 5.4 per cent, making 2.3 million tonnes more than in 1995.

- In China, in 1996, steel imports rose by 18.1 per cent to reach 26.7 million tonnes, while exports fell by nearly 50 per cent. Thus net steel imports went up by nearly 50 per cent.

- In the NIS, while steel imports changed little, they amounted to some 2.7 million tonnes in 1996; exports continued to rise by 3.4 million tonnes, or 12.5 per cent, to practically all parts of the world.

CRUDE STEEL PRODUCTION

- World: world crude steel production fell slightly in 1996 to 751.7 million tonnes, 0.5 per cent down on 1995.

- OECD: crude steel production for the area as a whole fell by 2.1 per cent in 1996, or 9.7 million tonnes less than in 1995. Output was of the order of 451 million tonnes.

- Crude steel production fell in all other Member countries, except Canada, Korea, Hungary and Mexico where output went up.

- With the exception of the African continent, including South Africa, where steel production fell in 1996, every other market economy area outside the OECD posted increases in steel production, by varying amounts, with a total increase of nearly 3 million tonnes.

- Steel production in Central and Eastern Europe fell during 1996. The situation was similar in the NIS where output continued to fall for all these republics, even though there was a recovery in steel production in Kazakhstan and Ukraine.

- In China, crude steel production reached a record level, exceeding 100 million tonnes for the first time. In 1996, China became the world's leading steel producer.

STEEL CAPACITY UTILISATION RATE

- In the OECD area as a whole, the average capacity utilisation rate fell slightly to 79 per cent in 1996.

- Capacity utilisation levels exceeded 90 per cent in Australia, Canada, Korea and the United States. Utilisation rates were 66 per cent in Japan and 74 per cent in the European Union.

- Steel production capacity continued to increase sharply in the other non-OECD market economy areas, but the average capacity utilisation rate fell appreciably to just under 70 per cent.

STEEL PRICES

- The fall in demand for steel in the OECD area and the continuation of stock rundowns generally led in 1996 to a surplus situation that did not lend itself to any increase in prices. Generally speaking the prices of flat products, like those of strip products, followed a downward trend throughout the first half of the year. However the price of strip tended to stabilise during the third quarter. The prices of flat products also steadied out a little later. Towards the end of the year a slight increase in steel prices could be discerned.

1997

APPARENT STEEL CONSUMPTION

- World: demand for steel should resume a steady increase at a little over 3 per cent, which would represent consumption more than 20 million tonnes up on 1996. 1997 would thus set a new yearly record with consumption of the order of 670 million tonnes.

- OECD: following the substantial fall in 1996, apparent consumption should rise by about 1.7 per cent, but at some 388.5 million tonnes would still be 2 per cent below the record level of 1995.

- With the exception of the United States, where consumption should remain steady but down on 1996, demand for steel is likely to increase in all the other Member countries. In Brazil, a substantial recovery of about 6 per cent in consumption is expected.

- Steel consumption should continue to rise in the Middle-East, India and the other countries of Asia, but is expected to fall in Latin America and South Africa. As regards the rest of the African continent, steel consumption should continue to decline for the fifth year in succession.

- In the central and eastern European countries, steel consumption is expected to rise even faster, possibly by more than 10 per cent.

- There should at last be a pick-up in demand for steel in the NIS, with apparent consumption increasing by nearly 19 per cent, or 5.7 million tonnes more than in 1996.

- In China, following the considerable increase in 1996, consumption is expected to rise by a little under 5 per cent, an additional 5.2 million tonnes.

STEEL TRADE

- World trade in steel should fall by about 3 per cent from the record levels achieved in 1995 and 1996.

- Total net exports from the OECD countries should continue to increase rapidly, by about 24 per cent, or a little over 5 million tonnes.

- Net exports from the European Union should fall as a result of the recovery in internal demand. Conversely in Japan a fall in imports and a slight increase in exports should bring about an increase of a little under 8 per cent in net exports.

- Due to the slowdown in domestic demand and the increase in output, net steel imports by the United States should fall by about 5.5 million tonnes, or a quarter less than in 1996.

- As regards the other market-economy areas outside the OECD, net steel imports should fall as a result of reduced imports and increased exports, particularly in Latin America and Asia.

- Net exports from the NIS are likely to fall by about 13.5 per cent, while exports from the different countries in this area should decline by at least 4 million tonnes from the exceptionally high level achieved in 1996.

- Net imports by China should continue their upward trend and could grow by about 18.7 per cent in 1997.

CRUDE STEEL PRODUCTION

- World: like consumption, world crude steel production should increase by about 3.1 per cent in 1997, or about 23 million tonnes, reaching some 775 million tonnes.

- OECD: crude steel production in the OECD area is likely to go up by about 2.9 per cent, representing nearly 13 million tonnes more than in 1996. Output should increase in practically all Member countries.

- In Brazil, production should continue to increase by about 0.5 million tonnes, or a little over 2 per cent.

– In all the other market-economy areas, again excluding the African continent, steel production should continue to increase by about 6.5 per cent, representing an additional 5 million tonnes.

– As regards the NIS, output should increase by about 3 per cent, or 2.4 million tones, as a result of the rises particularly in Ukraine and Kazakhstan.

– Production in China should continue to rise, but less rapidly, by about 1.3 per cent or 1.4 million tonnes more than in 1996, and China is likely to remain the world's leading steel producer.

CRUDE STEEL CAPACITY UTILISATION

– Crude steel production capacity in the OECD area as a whole can be expected to increase by 21 million tonnes in 1997. The average capacity utilisation rate is expected to fall from 79 per cent in 1996 to 78 per cent in 1997 as a result of this increase, and despite the expected rise in output.

– In the remaining market economies, crude steel production capacity should grow by a further 25 million tonnes. As a result, the capacity utilisation rate is expected to fall to no more than 60 per cent.

STEEL PRICES

– With the rundown in stocks coming to an end in 1996, and the recovery in demand for steel, the increase in the prices of steel products which began at the end of 1996 should speed up under the favourable effect of the recovery in the market, the need to build up stocks and the reduction in import pressure.

1998

APPARENT STEEL CONSUMPTION

– World: the recovery in world demand for steel that took place in 1997 is expected to continue in 1998 under the impetus of certain cyclical effects, particularly if the current forecasts of economic growth, which should speed up in 1998, come to pass. The increase in steel consumption could be about 3.5 per cent, which would amount to some 24 million tonnes more than in 1997.

– OECD: with economic growth for the zone as a whole expected to be of the order of 2.7 per cent, the increase in apparent steel consumption could reach 3.75 per cent for the whole of the OECD area. This increase would represent some 14.6 million tonnes and takes into account some probable restocking in a number of Member countries.

– All Member countries should benefit from this trend, even though the rises in apparent steel consumption are expected to be greater in Europe and Mexico than in the United States or Japan.

– In Brazil, demand for steel is expected to rise substantially, by about 14.5 per cent, or nearly 2 million tonnes more than in 1997.

– In the non-OECD market economies taken overall, the increase in steel consumption should accelerate and reach about 3.8 per cent, which would represent an increase of about 4.5 million tonnes. The biggest rises would be in Latin America, the Middle East and India.

– In the NIS, although steel consumption is beginning to increase in 1997, it is expected to continue in 1998 at a rate between 8.5 and 9 per cent, which would account for a little over 3 million tonnes. It should, however, be noted that at a little under 39 million tonnes, this level of consumption still represents only one-third of the 1988 figure.

– In China, demand for steel could fall slightly by 0.6 per cent, making 0.7 million tonnes less than in 1997.

Main results

	1996 variation in		1997 variation in		1998 variation in	
	Million tonnes	%	Million tonnes	%	Million tonnes	%
a) Change in apparent steel consumption (in product equivalent)						
United States	+6.0	+6.2	−3.1	−3.0	+1.1	+1.1
Japan	−1.0	−1.2	+0.5	+0.6	+2.0	+2.5
EU (15)	−16.0	−12.6	+5.6	+5.0	+6.6	+5.7
Other Europe	−4.7	−16.3	+1.4	+5.6	+2.5	+9.9
Canada	−0.4	−2.8	+0.1	+0.4	+0.2	+1.7
Korea	+1.3	+3.7	+1.1	+2.8	+1.1	+2.9
Mexico	+0.5	+6.3	+0.9	+10.6	+1.0	+10.0
Oceania	−0.2	−3.6	+0.2	+2.5	+0.2	+2.9
Total OECD	−14.4	−3.6	+6.6	+1.7	+14.6	+3.8
Brazil	−0.1	−0.6	+0.7	+6.0	+1.9	+14.6
Other market economies	+5.2	+4.8	+1.6	+1.4	+4.5	+3.8
Central and Eastern Europe	+0.2	+5.6	+0.5	+10.8	+0.3	+5.8
NIS	−3.2	−9.6	+5.7	+18.9	+3.1	+8.7
China and North Korea	+10.8	+11.2	+5.2	+4.9	−0.7	−0.6
World	−1.5	−0.2	+20.2	+3.1	+23.7	+3.5
b) Net trade (in product equivalent) See section *d)* below for detailed import and export figures						
EU (15) net exports	+8.5	+68.5	−2.6	−12.4	−1.5	−8.2
Japan net exports	−1.6	−10.8	+1.1	+7.9	−0.6	−3.9
United States net imports	+6.2	+39.3	−5.5	−25.3	−2.2	−13.3
Other OECD net exports	+6.3	+271.7	+1.2	+13.9	+0.5	+5.0
Total OECD exportations	+6.9	+49.6	+5.3	+25.5	+0.5	+1.9
Other market economies net imports	+3.7	+8.7	−4.2	−9.1	+1.6	+3.8
China net imports	+6.7	+51.1	+3.7	+18.6	−2.5	−10.6
c) Production of crude steel						
United States	−0.5	−0.5	+2.8	+3.0	+3.8	+3.9
Japan	−2.8	−2.8	+1.6	+1.7	+1.5	+1.5
EU (15)	−8.6	−5.5	+3.3	+2.3	+5.6	+3.7
Other Europe	−1.0	−2.9	+1.1	+3.2	+3.0	+8.4
Canada	+0.2	+1.5	+1.0	+6.6	+0.4	+2.6
Korea	+2.1	+5.8	+2.2	+5.7	+2.1	+5.0
Mexico	+1.0	+8.4	+0.7	+5.5	+0.4	+3.1
Oceania	−0.1	−1.0	+0.1	+1.1	+0.1	+1.5
Total OECD	−9.7	−2.1	+12.9	+2.9	+17.0	+3.7
Brazil	+0.2	+0.7	+0.5	+2.1	+1.1	+4.3
Other market economies	+2.8	+3.6	+5.1	+6.5	+3.2	+3.8
Central and Eastern Europe	−1.1	−8.4	+0.8	+6.5	+0.7	+5.3
NIS	−0.8	−1.1	+2.4	+3.0	+3.2	+4.0
China and North Korea	+5.0	+4.9	+1.4	+1.3	+1.3	+1.2
World	−3.7	−0.5	+23.1	+3.1	+26.4	+3.4

	1996			1997			1998		
	Imports	Exports	Net trade	Imports	Exports	Net trade	Imports	Exports	Net trade
d) Imports, exports and net trade (in million tonnes)									
EU (15)	13.0	33.9	−20.9	13.2	31.4	−18.3	12.0	28.7	−16.8
Japan	5.9	19.3	−13.3	5.1	19.5	−14.4	5.5	19.3	−13.8
United States	26.5	4.6	21.9	21.7	5.3	16.4	20.2	6.0	14.2
Rest OECD	28.0	36.4	−8.4	26.1	36.0	−9.9	25.0	35.3	−10.3
Total OECD	73.3	94.2	−20.9	66.2	92.2	−26.0	62.7	89.3	−26.6
Other market economies	61.4	26.6	34.8	60.8	28.6	32.2	63.7	28.7	34.7
NIS	2.7	30.7	−28.0	2.3	26.5	−24.2	2.5	26.0	−23.5
Central and Eastern Europe	1.5	6.7	−5.2	1.6	6.9	−5.4	2.0	7.6	−5.7
China and North Korea	26.7	6.9	19.8	30.0	6.5	23.5	29.0	8.0	21.0
World	165.6	164.9	0.7	160.7	160.5	0.1	159.6	159.6	0.0

Main results *(cont.)*

	Capacity in million tonnes			Utilisation rate in %		
	1996	1997	1998	1996	1997	1998
	e) Crude steel capacity[1] and utilisation rates					
United States	105.2	109.3	110.2	90	89	92
Canada	15.5	16.7	16.7	94	93	96
Korea	42.9	45.8	49.2	91	90	88
Mexico	15.2	16.3	16.3	87	85	88
Oceania	9.7	9.8	9.8	95	95	96
Japan	149.7	149.7	149.7	66	67	68
EU (15)	198.8	200.3	202.2	74	75	77
Other Europe	43.3	43.7	44.1	80	82	88
Total OECD	570.6	591.6	598.2	79	78	80
Brazil	31.1	32.5	33.9	81	79	79
Other market economies	114.7	139.5	173.2	69	60	50
Total market economies	716.4	763.6	805.3	85	82	81

1. Estimates of capacity for OECD countries are based on the Steel Committee's Annual Survey of Effective Capacity. There are differences in country definitions and changes in a country's operating rates from year to year are more significant than direct comparisons between the various countries' operating rates.

Source: OECD Secretariat.

STEEL TRADE

– World trade in steel could again fall slightly, by about 0.7 per cent, and would represent no more than 23 per cent of the world steel market.

– Total net exports from the OECD area could rise by a further 2.3 per cent, this slight increase being due primarily to a bigger fall in imports than in exports. Net exports from EU (15) should fall by a little over 8 per cent. Net exports from Japan could also decline by a little less than 4 per cent. As a result of the commissioning of new capacity, and a slight rise in exports, net imports by the United States could fall by 13.3 per cent.

– Net imports by the other market economies outside the OECD area are likely to continue rising at a steady rate of about 9 per cent, or more than an additional 2 million tonnes. It would primarily be the net imports of Asia and the Middle East that would increase, while India could start to become a net steel exporter.

– Net exports from the NIS could continue to fall. This would be the combination of a fall in exports and a slight recovery in imports.

– In China, net imports of steel could fall by nearly 10 per cent.

CRUDE STEEL PRODUCTION

– World: with the acceleration of world steel consumption, crude steel production could increase by a little over 26 million tonnes, or nearly 3.5 per cent. Total world steel production could exceed 800 million tonnes.

– OECD: steel production in the area as a whole could increase by nearly 3.7 per cent, representing some 17 million tonnes.

– Growth of production is likely to vary across Member countries, being higher in the United States, Europe, Korea and Mexico, with smaller increases in Japan and Australia.

– In Brazil, crude steel production should continue to rise by about 4.25 per cent, representing an additional amount of just over a million tonnes.

- The upward trend in steel production should also continue in the other non-OECD market-economy areas with the increase totalling about 3.75 per cent. In the NIS, there could also be a rise of some 4 per cent. In China the rise in production is likely to be somewhat less, around 1.2 per cent.

STEEL PRODUCTION CAPACITY UTILISATION

- In 1998, crude steel production capacity in the OECD area as a whole will increase by a further 6.6 million tonnes. Taking account of the sharp rise in production, the average capacity utilisation rate should be around 80 per cent.

- In the other market-economy areas, steel production capacity should continue to increase, by nearly 24 per cent, which would be equivalent to an additional 34 million tonnes. The additional capacity of 1997 and 1998 is unlikely to be operational by the planned commissioning date, and consequently the capacity utilisation rate in all these areas could fall to no more than 50 per cent.

STEEL PRICES

- As forecast, steel market trends in 1998 are likely to result in an increase in the price of steel products, which will probably be more pronounced for flat products than for long products.

3

DEVELOPMENTS IN THE STEEL MARKET BY AREA

UNITED STATES

Economic growth in the United States in 1996 continued to surpass predictions. GDP, which rose by 2.4 per cent over the year, even went up by 3.9 per cent in the fourth quarter after a slight slowdown to 2.1 per cent in the third quarter. This spurt at the end of the year was the result of the increase in private consumption, as consumers gained confidence in the American economy, and of an increase in disposable incomes. Unemployment continued to fall, and more than 2.6 million jobs were created during the year. Despite the healthy economy and an historically low unemployment rate, inflation remained under control, with producer prices and consumer prices not rising by more than 2.5 per cent over the year.

Industrial output went up by 2.7 per cent over the year, the sectors performing best being electrical and non-electrical machinery. The automobile industry went against this trend, with output falling by 4 per cent compared with 1995. The construction sector was particularly dynamic, especially as regards house building, where housing starts increased by 8 per cent. Non-residential construction was up by 3 per cent. The principal negative feature of the American economy in 1996 will have been the rise in its trade deficit which, at 114 billion dollars, will have reached its highest level for eight years.

Economic growth in 1997 should slow down somewhat and GDP is expected to grow by about 2.2 per cent. After all, the main factors that boosted growth in 1996 had not been expected to continue active throughout 1997. However inflation should remain under control, probably around the 1996 values. Job creation is expected to continue at a steady rate and unemployment could stabilise at the 1996 level. Growth in investment is likely to be less and total domestic demand should slow down, rising by only 2.1 per cent. A strong dollar could result in the substantial trade deficit continuing.

Industrial production can be expected to slow in relation to the 1996, increasing by no more than 2.6 per cent. A rather lethargic first six months could be followed by a more vigorous second half of the year. The automobile industry should resume growth with a level of at least 1 to 2 per cent being expected. Activity in residential construction could fall by 3 per cent, while in non-residential construction it should continue to rise at the same rate as in 1996.

The forecasts currently available suggest that growth of the American economy in 1988 should revert to its long-term trend and thus reach about 2 per cent. Fiscal stringency is likely to persist, and activity in those sectors sensitive to long-term interest rates could decline somewhat. Unemployment is expected to change little by comparison with 1996 and 1997, with private consumption rising as in 1997 by about 2.2 per cent, investment growing less sharply, but industrial output continuing to grow by 2.5 per cent.

The growth in the overall economy in 1996 was beneficial to the steel market. Apparent consumption expressed in finished product equivalents was 6.2 per cent up on 1995 at 103.5 million tonnes, its highest level since 1978. American steelmakers' deliveries to the domestic market rose by 3.1 per cent to 91.2 million tonnes. As far as the main consumer sectors were concerned, steel deliveries to the construction sector rose by 10.2 per cent and those to the automobile sector by 6.6 per cent. In other industries, deliveries also went up substantially, by 9.1 per cent for the energy sector and 12.3 per cent for the agricultural machinery sector. Deliveries to steel stockholders, which account for one-quarter of total deliveries, went up by 7.7 per cent.

Crude steel production fell by 0.5 per cent to 94.7 million tonnes. This fall was partly due to problems at a number of blast furnaces and to a strike. Following an increase of 3.1 million tonnes in capacity, the capacity utilisation rate fell slightly to 90 per cent in 1996. The proportion of continuously cast steel again

Production indices, 1990 = 100 (seasonally adjusted)

	1994	1995	1996	1996 Fourth quarter	1996/95 % change
Industrial production	109.8	113.4	116.5	118.4	2.7
Manufacturing industries	111.1	115.0	118.1	120.3	2.7
Motor vehicles and parts	136.1	138.9	133.4	132.7	−4.0
Fabricated metal products	110.6	114.3	117.1	117.9	2.4
Non-electrical machinery	125.1	141.1	156.2	161.3	10.7
Electrical machinery	143.9	168.9	186.2	190.9	10.2
Mining	98.0	97.3	99.3	100.2	2.1

Source: OECD, *Indicators of Industrial Activities.*

increased by 2 per cent to 93 per cent. Prices of steel products fell on average by 3.6 per cent in 1996 compared with 1995, and the profitability of the steel industry declined over the year as a whole.

As regards trade in steel products, heavy demand led to an increase of 19.5 per cent in imports, or 4.3 million tonnes more than in 1995, even though imports fell during the first half of the year. The rate of import penetration on the American market went up from 22.6 in 1995 to 25.9 in 1996. Meanwhile steel exports fell by 29 per cent from the record level they had reached in 1995.

As a result of the shutdown of a number of blast furnaces, imports of semi-finished products rose by 55 per cent to 6.8 million tonnes. Other products for which imports rose strongly were heavy sections (+54 per cent), plate (+38 per cent), and hot rolled sheet (+29 per cent). Imports from EU (15) rose by 48 per cent to 8 million tonnes, and those from Mexico by 28 per cent. Imports from Brazil and Russia also rose sharply. Conversely, imports from Japan and Korea fell by 19 and 2 per cent respectively. Steel exports to Mexico rose by a little over 5.5 per cent while those to Canada fell slightly. Exports to the EU (15) fell by 61 per cent, as did those to South-East Asia, especially Korea and Chinese Taipei, which were down by 80 and 70 per cent respectively.

With the expected economic growth in 1997 and the favourable outlook for the main steel consuming industries, demand for steel is expected to remain high. However apparent consumption could fall by about 3 per cent, due mainly to the build-up of stocks during 1996. Imports are expected to fall appreciably, owing to the recovery in consumption on the markets of the main partner countries and the introduction of new capacity, with some 3.9 million tonnes expected for 1997. Imports of semi-finished products should show a particularly marked fall. The additional capacity should not only enable American producers to respond better to the demand from the domestic market but should also increase exports. Crude steel production should rise by about 3 per cent to 97.5 million tonnes.

In 1998, the market should remain healthy with apparent consumption rising by a little over 1 per cent, assisted by stock changes. Crude steel production could rise by 3.9 per cent to over 101 million tonnes. Despite the commissioning of about 1 million tonnes of additional capacity, the capacity utilisation rate would rise to 92 per cent. Imports are likely to continue to decline, with exports rising by more than 10 per cent.

CANADA

Economic growth in Canada slowed down in 1996 with GDP going up by only 1.4 per cent. This slowdown is explained in part by a considerable rundown in stocks. However improved public finances and the better competitive position enabled the economy to show a marked recovery during the second half of the year. A fall in the exchange rate and in short-term interest rates allowed investment and private consumption to recover. Inflation slowed down to no more than 1.6 per cent in 1996. Unemployment rose slightly to 9.7 per cent in 1996 compared with 9.5 per cent in 1995.

As industry ran down stocks, industrial output increased by only 1.4 per cent. However towards the end of the year there was an upturn to some 4.2 per cent in the fourth quarter. In the automobile sector, production in 1996 fell by 1.1 per cent, while sales rose by 3.3 per cent. Activity in the construction sector rose by 6.4 per cent over the year, with a sharp increase of 18.9 per cent as regards residential construction and a fall of 8.1 per cent for non-residential construction. Pipe-making increased by 10.3 per cent over the year.

Crude steel production in Canada was up 1.5 per cent in 1996 to 14.6 million tonnes, 97.6 per cent of which was made by continuous casting. Steel imports remained high, partly as a result of imports of semi-finished products which Canadian producers needed to meet demand, but were nevertheless 14 per cent down on 1995 at 4.5 million tonnes. Exports also fell, but only by 3.5 per cent. Apparent consumption was down by 2.8 per cent at 0.4 million tonnes less than in 1995, again partly due to reduction in stocks. The prices of steel products fell in Canada in 1996 and although output was up, steelmakers' incomes fell.

Projections for Canadian economic activity in 1996 and 1997 forecast a clear recovery and GDP is expected to rise by +3.2 per cent in 1997 and +3 per cent in 1998. Inflation in 1997 may remain at the 1996 level, *i.e.* 1.6 per cent, and increase slightly to 1.9 per cent in 1998. Unemployment should fall again to 9.5 per cent in 1997 and is likely to fall to 8.9 per cent in 1998. Private consumption is expected to grow faster, together with investment.

In 1997, industrial output is likely to go up by 4 per cent. Production of motor vehicles should rise by 4.7 per cent, with sales of new vehicles increasing by the same proportion, assisted by relatively low interest rates. In the construction sector, too, interest rates should permit project starts to go up by more than 10 per cent. A similar trend is expected for 1998, with industrial output rising by about 3.9 per cent with an increase of more than 8 per cent in activity in the residential construction sector.

Apparent steel consumption in Canada should begin to pick up and increase by about 0.4 per cent in 1997. This modest recovery stems from the expected fall of 22 per cent in imports, particularly as regards semi-finished products. Exports could fall by 4.5 per cent. Crude steel production, as new facilities are commissioned, is likely to go up by 6.6 per cent, or 1 million tonnes more than in 1996. In 1998, steel production could rise by a further 1.7 per cent. Crude steel output should rise by 2.6 per cent, with imports continuing to fall by nearly 9 per cent.

MEXICO

Following the deep recession experienced by Mexico in 1995, renewed economic growth was a feature of 1996 with GDP going up by 4 per cent. Investment rose by more than 8 per cent and private consumption was also on an upward trend. The economic recovery was also underpinned by a substantial increase in imports. As a result of the considerable readjustment in the exchange rate in 1995 and the progress made in stabilising the economy, a possible upturn in economic growth is predicted in 1997. GDP could rise by 5 per cent with increased confidence generating an increase in domestic demand of about 4.6 per cent and in investment of 10 per cent. This trend could also continue in 1998.

Following the substantial fall in steel consumption in 1995, the demand for steel went up by 6.3 per cent in 1996, or 500 000 tonnes more than in the previous year. Steel exports remained at a high level, at more than 2.5 million tonnes. Imports also began to recover, going up by 10 per cent. Crude steel production rose by 8.4 per cent, or 1 million tonnes more than in 1995, and exceeded 13 million tonnes, of which more than 80 per cent was produced by continuous casting.

In 1997, steel consumption is expected to grow more quickly, rising by more than 10.5 per cent to 9.8 million tonnes, or 1 million tonnes more than in 1996. Steel imports should stay at 1996 levels while exports, affected by the recovery in domestic demand, could fall. Crude steel production should go up by 5.5 per cent, or 0.7 million tonnes more than in 1996. In 1998, steel consumption should again rise sharply, by about 10 per cent, or more than 1 million tonnes higher than in 1997. Steel imports are expected to remain steady at 0.7 million tonnes with exports falling by a further 27 per cent approximately. Crude steel production should increase by about 3 per cent with the additional 0.4 million tonnes bringing the total to 14.3 million tonnes.

EUROPEAN UNION (15)

The 1995 slowdown in the European economy changed into an upturn by the middle of 1996. Confidence in the industry improved and order books filled up. Consumer confidence steadied out as from the beginning of 1996. Real growth in GDP in the EU area as a whole was 1.6 per cent. However, GDP growth was below the EU mean in Germany, France, Italy, Austria and Belgium.

The state of the labour market remained unsatisfactory, and unemployment in the Community as a whole was slightly under 11 per cent in 1996. Inflation, on the other hand, remained under control and did not exceed 2.6 per cent. Private consumption went up slightly, being 1.9 per cent above that of 1995. Private non-residential investment increased by 3.5 per cent, while any growth in private investment in the housebuilding sector remained at a very low level. Industrial output stagnated throughout the EU at the previous year's level, and in a number of Member States actually fell, by 0.9 per cent in Germany, 2.8 per cent in Italy and 0.7 per cent in Spain. Although the motor industry saw business expand by 2.6 per cent, clear recessional signs appeared in the construction sector. This downward trend was more marked in the housebuilding sector, while activities in the non-residential construction sector appeared to steady out in 1996.

It appears that the gradual upturn in the EU economies from the end of 1996 should continue throughout 1997, with mean GDP expected to increase by about 2.4 per cent. This growth seems due primarily to exports, which are benefiting from favourable foreign demand and to an improvement in the competitivity of European products as a result not only of the value of the dollar but also of cost control in Europe. Private consumption is expected to continue supporting activity in 1997 with investment once again becoming the prime mover of this growth. In 1997, unemployment is expected to decline, with inflation – averaging 2.2 per cent – brought under control. As regards the steel-consuming sectors, construction is expected to remain depressed, although with a degree of stabilisation that could show up in residential construction. Again the levelling-off in non-residential construction seen in 1996 could give way to slight growth. Mechanical engineering, which has been expanding for a number of years, is expected to maintain this trend in 1997, even though the rate of growth could slow down somewhat by comparison with 1996, with the automobile industry staying on the same trend of medium growth.

For 1998, there should be an upturn in economic growth within the EU, with GDP rising by 2.8 per cent. Similar developments are expected in private consumption and investment. Unemployment should continue to fall to 10.3 per cent of the working population.

In the steel sector, the EU (15)'s apparent consumption, as measured by finished product equivalents, appeared to have fallen by 12.6 per cent in 1996, representing some 16.0 million tonnes less than in 1995.

Production indices, 1990 = 100 (seasonally adjusted)

	1994	1995	1996	1996 Fourth quarter	1996/95 % change
Industrial production					
Germany	97.4	98.1	97.2	97.9	−0.9
Spain	98.7	103.3	102.6	104.1	−0.7
France	97.5	99.3	99.7	99.8	0.4
Italy	101.7	107.8	104.8	103.8	−2.8
United Kingdom	103.4	105.9	107.6	108.1	+1.6
EU (15)	100.1	103.4	103.4	103.9	−
Metal-using industries:					
EU (15)	99.4	103.2	103.5	104.0	+0.3
of which:					
motor vehicles	94.5	99.7	102.2	103.0	+2.6
mechanical engineering	88.7	96.6	96.7	95.7	0.0

Source: EUROSTAT, *Panorama of European Industry;* OECD, *Indicators of Industrial Activity.*

**Yearly percentage changes in real
and apparent steel consumption in the European Union area**

	1996/95 Realised		1997/96 Estimates		1998/97 Forecast	
	Real	Apparent	Real	Apparent	Real	Apparent
Germany	−12.6	−15.9	+13.2	+14.3	+4.0	+3.7
Spain	−8.8	−17.6	−2.6	+6.5	+10.8	+7.8
France	+4.2	−8.6	0.0	+0.7	+1.4	+6.2
Italy	+3.1	−9.2	−14.2	−7.3	+10.1	+8.4
United Kingdom	−0.8	−5.8	+6.1	+8.4	+2.9	+2.1
Rest EU (15)	−3.1	−13.8	−1.4	+1.9	+1.4	+6.2
Total EU (15)	−1.7	−12.6	−2.1	+5.0	+5.2	+5.7

Source: OECD Secretariat.

The extent of stock drawdowns, both by producers and stockholders, is largely responsible for this fall. Real steel consumption in the EU apparently fell by only 1.7 per cent.

Crude steel production in the EU as a whole fell by 8.6 million tonnes in 1996, or 5.5 per cent, to 147.0 million tonnes. Output fell by varying amounts in most countries of the Community, except for Denmark, Finland, Ireland, Portugal and the United Kingdom. The sharp fall in domestic demand led to a 35 per cent drop in steel imports, while over the same period exports rose by a little over 4.4 per cent. Despite the fall in steel production, the market nevertheless showed a surplus, resulting in an appreciable fall in prices. During the first half of 1996, prices of ordinary steel flat products fell by 10 to 20 per cent and those of strip products by a further 10 per cent, thus once again returning to very low levels. However prices steadied during the third quarter of 1996 for long products and a little later in the year for strip products, which even showed a certain upturn.

In 1997, crude steel production in the EU is expected to rise by about 2.3 per cent, an increase of 3.3 million tonnes. The capacity utilisation rate should increase to 75 per cent. Imports of steel are expected to increase by 1.3 per cent compared to 1996 levels while exports should rise by 4.4 per cent. There could be a resumption of stock build-ups and apparent consumption could go up by 5.1 per cent, or some 5.7 million tonnes more than in 1996. The rise in the price of steel products that began at the end of 1996 should speed up.

In 1998, apparent consumption of steel should steadily rise further, by 5.7 per cent, or 6.6 million tonnes more than in 1997. Inventories are expected to be reconstituted, particularly by traders and consumers. Owing to the rising demand on the domestic market, steel exports should fall still further by about 8.5 per cent. As a result, crude steel output could increase by 3.7 per cent, representing 5.6 million tonnes more than in 1997, bringing total EU production to a level above that achieved in 1995.

OTHER EUROPEAN COUNTRIES

Besides Norway, Switzerland, Turkey and the former Yugoslavia, this area now includes three new countries, Hungary, Poland and the Czech Republic, which have joined the OECD since 1996. Steel consumption in the area as a whole fell by 16.3 per cent in 1996, a reduction of 4.7 million tonnes compared with 1995. The largest decline was reported in Turkey. Crude steel production was down slightly by 2.9 per cent, a reduction of 1 million tonnes. Imports declined by 15.8 per cent, but exports were up by 17.1 per cent.

In 1997, steel consumption should rise by 5.6 per cent, an increase of 1.4 million tonnes, and imports should be up by 3.6 per cent. Exports should also rise slightly, by 1.2 per cent, which should lead to an increase of 3.2 per cent in crude steel output, thus bringing it back to the 1995 level.

In 1998, growth in steel consumption should accelerate to around 9.9 per cent, an increase of 2.5 million tonnes. This increase in volume terms will probably be driven by higher output from Poland

and the former Yugoslavia. Net trade in steel products will probably remain relatively unchanged compared with 1997, but there should be strong growth of almost 8.5 per cent, equivalent to 3 million tonnes, in crude steel production.

In 1996, **Norway's** GDP grew at a rate of just over 5 per cent, buoyed in part by high output from the oil sector, and outpaced that of most OECD Member countries. Growth in domestic demand was proportionately lower, however, at around 4 per cent. The on-shore economy is expected to slow to around 3 per cent in 1997, and a further slowdown in 1998 might reduce it to no more than 2.3 per cent. Domestic demand is also expected to slow, declining to 2.7 per cent in 1997 and to 2.4 per cent in 1998.

Steel production in Norway fell off in 1996, particularly during the second half of the year, and declined by a total of 2 per cent. Apparent steel consumption seems to have fallen by 5.8 per cent over the year, primarily due to reduced investment in the construction of off-shore platforms and modest growth in the building sector. A 2 to 3 per cent decline is projected in real steel consumption in 1997, although apparent consumption should remain at the 1996 level. Crude steel production is expected to rise by 10 per cent. 1998 should see a small 0.8 per cent increase in apparent steel consumption, while steel production is expected subside to the 1996 level.

In **Switzerland**, GDP declined by 0.7 per cent in 1996. New orders to industry continued to fall, except in the case of major exporting firms. Household consumption grew by a mere 0.3 per cent as a result of the decline in real disposable income. Investment in the construction sector, in both the public and private sectors, was down by around 3.5 per cent in 1995. Investment in capital goods was the only component of domestic demand in which there was significant growth, amounting to around 5 per cent.

Apparent steel consumption fell by 15.8 per cent in 1996. In contrast, crude steel production was 4.8 per cent up on 1995. This increase was due to the fact that, unlike in the previous year, production was not shut down for the customary two months of maintenance and upgrading work. Production facilities were only used at 81 per cent of capacity. With regard to foreign trade, imports of steel products were down by around 21 per cent, falling from 1.99 million tonnes in 1995 to 1.57 million tonnes in 1996. Exports also declined by 11 per cent to no more than 0.74 million tonnes. The traditional deficit in the balance of trade for steel products therefore fell by 29 per cent in 1996.

With growth in GDP expected to amount to a mere 0.5 per cent in 1997, the Swiss economy will again perform poorly. The downward trend in the construction sector is likely to persist. The decline in investment in this sector is expected to accelerate, rising to 4 per cent. Unemployment is expected to pass the 5 per cent mark. The strength of the Swiss franc will probably continue to be a problem for Swiss steelmakers. Apparent steel consumption should fall still further and crude steel output may well be down by almost 13 per cent.

From 1998 onwards, the impact on the Swiss economy of a strong Swiss franc should start to ease. GDP is expected to grow by around 2 per cent. Demand from the construction sector may well pick up once current stocks have been drawn down. Private consumption and investment may start to recover, growing by 1.5 and 1.9 per cent respectively. Industrial output should grow by around 4 per cent. Unemployment should subside to 4.5 per cent. Apparent steel consumption should start to recover, rising by almost 9 per cent. As a result, crude steel production should increase by 21.5 per cent.

The sharp recovery in the **Turkish** economy reported in 1995 gathered pace in early 1996 and during the first six months of the year GDP grew at a rate of about 10 per cent. The GDP growth rate for the year as a whole amounted to some 7.5 per cent. Domestic demand rose substantially, as did investment and household consumption. Apparent steel consumption fell by 24.4 per cent, however, probably as a result of the substantial build-up of stocks in 1995. While crude steel production, at 13.4 million tonnes, was up by almost 5 per cent, steel imports fell by over 30 per cent, despite a 53 per cent increase in exports over the same period.

1997 and 1998 should see the growth in the economy stabilise and GDP should grow by 5.7 per cent in 1997 and by a further 5 per cent in 1998. Industrial output should continue to rise, growing by 7 per cent in 1997 and by a further 6 per cent in 1998. Apparent steel consumption should rise by almost 3 per cent in 1997 and remain at that level in 1998. Crude steel production should be up by over 9 per cent in 1997 and by a further 3.5 per cent in 1998, when it should pass the 15 million tonne mark. Steel imports, on the other hand, will probably decline over the two years, while exports should continue along an upwards trend.

While the impacts of the stabilisation programme in **Hungary** continued to make themselves felt in 1996, restrictive economic policies precluded a return to growth. GDP grew by merely 0.5 per cent. The rate of inflation declined, as did wages and real disposable income. Investment remained stagnant. Industrial output rose, but only by 2.8 per cent. The economic recovery should start to gather pace in 1997, with GDP growing by 1.7 to 2.0 per cent, and should start to accelerate in 1998 which should see growth of the order of over 3 per cent. This should be matched by growth in industrial output of 5 per cent in 1997 and 6 per cent in 1998. The same trend should emerge in investments, which should grow at an even faster pace.

1996 was therefore marked by a decline of almost 12.8 per cent in steel consumption. Steel production, which had plummeted in 1995 as a result of restructuring in the steel industry, remained stagnant in 1996, growing by merely 0.5 per cent. While steel exports were up slightly, imports fell sharply by 21.6 per cent. Steel consumption should rise by 2.4 per cent in 1997. Production of crude steel will probably remain stagnant at 1996 levels, but imports should continue to rise and should be up by 10 per cent. Steel consumption should increase by 3.2 per cent in 1998, while crude steel production should continue to rise and should be up by about 8 per cent.

The **Polish** economy continued to grow strongly for the fifth year running. GDP grew by 5.5 per cent, private consumption by 7 per cent and investment by 21 per cent. The rate of unemployment fell from 13.3 per cent in 1995 to 12.5 per cent in 1996. Industrial output rose by 8 per cent and, in terms of steel-using sectors, the highest rates of growth were achieved in the car manufacturing and shipbuilding industries. In 1997 and 1998, the economy should grow at a rate of over 5 per cent. Industrial output should rise by 7.5 per cent in 1997 and by a further 7 per cent in 1998. Unemployment should continue to fall steadily. Inflation is forecast to remain below 13 per cent in 1997 and should fall by a further two percentage points in 1998.

Despite this satisfactory economic performance, for no clear reason apparent steel consumption fell by 13 per cent in 1996. Crude steel production similarly fell by 12.3 per cent. Steel imports and exports rose by 16 and 4.3 per cent respectively. Steel consumption looks set to remain stagnant in 1997 at the 1996 level. It should then resume an upward trend in 1998 and grow strongly since, after two years of stabilisation, consumption is expected to rise by 14 per cent. Crude steel production may well fall by a further 3.4 per cent in 1997, followed by growth of over 10 per cent in 1998. To compensate for this trend, steel imports should increase by 5.9 per cent in 1997 and then drop by 20 per cent in 1998. Exports are expected to decline by around 3 per cent over these two years.

The economy of the **Czech Republic** grew by 4.4 per cent in 1996, although this rate of growth might gradually start to accelerate over the next two years. The rate of unemployment remained extremely low at around 3 per cent in 1996, although a slight increase is possible over the following two years. While growth in industrial output slowed from 9.2 per cent in 1995 to 6.8 per cent in 1996, it should recover in 1997 to 7.5 per cent, rising to 8.5 per cent in 1998. Output has primarily slowed in the construction sector, where growth amounted to no more than 4 per cent in 1996.

The privatisation of the Czech steel industry has now entered its final stage. Apparent steel consumption fell by 13.2 per cent in 1996 and may well remain stagnant at that level in 1997 before rising by over 20 per cent in 1998. Trade in steel is not expected to vary significantly. Crude steel production, which had fallen by 9.2 per cent in 1996, may well decline by a further 3.4 per cent in 1997 before rising by 12.7 per cent in 1998.

JAPAN

The Japanese economy continued to recover in 1996. Growth in GDP, following rapid growth of 6.5 per cent during the first half of the year, should amount to around 3.5 per cent in 1996. Higher incomes and strong growth in industrial investment have cancelled out the adverse impact of restrictive fiscal policies on demand. Public investment was up sharply, by 13.6 per cent, as was private residential investment which rose by 12.7 per cent.

Indices of activity in steel-consuming sectors, 1990 = 100

	1994	1995	1996	1996/95 % change
Industrial production	93.1	96.2	98.5	+2.4
Production of passenger cars	86.5	84.4	85.2	+0.9
Production of commercial vehicles	88.2	86.8	87.8	+1.2
Non-electrical machinery	76.7	81.4	86.0	+5.7
Electrical machinery	104.0	116.1	124.1	+7.0
Shipbuilding	106.6	112.9	116.6	+3.3

Source: OECD, Indicators of Industrial Activity.

Industrial output rose by 2.4 per cent in 1996. After a decline in activity in the car manufacturing and commercial vehicle sectors in 1995, both sectors recovered in 1996 with a 0.9 per cent increase in the car sector and a 1.2 per cent increase in the commercial vehicle sector. Output in most other industrial sectors steadily gained ground.

Over the next few years, the Japanese government intends to strengthen and consolidate the expansion of the economy in order to ensure stable medium- and long-term growth. At the same time, the government will press ahead with deregulation and the promotion of other reforms to restructure the economy. Fiscal consolidation in 1997 should limit growth to around 1.6 per cent, although growth should pick up in 1998 to 3.7 per cent. Industrial output should follow an upward trend and is expected to rise to 3.5 per cent in 1997 and 4.7 per cent in 1998.

In the steel sector, apparent consumption in Japan fell to 1.2 per cent in 1996, a million tonnes less than in 1995. Stock adjustment, which was completed by the autumn of 1996, meant that real steel consumption fell by only 0.4 per cent in 1996. Steel consumption should begin to recover in 1997 and should rise by 0.6 per cent followed by stronger growth of around 2.5 per cent in 1998.

The levelling-off of steel demand in 1996, coupled with stock reduction by both consumers and merchants, would appear to have contributed to a 2.8 per cent decline in crude steel production which, at 98.8 million tonnes, was 2.8 million tonnes down on 1995 levels. The capacity utilisation rate amounted to 66 per cent. Most Japanese producers have reported higher earnings, primarily as a result of cost-cutting exercises and a weak yen. In 1997, renewed demand and the relatively low level of stocks should pave the way for a 1.7 per cent increase in crude steel production which, at 1.6 million additional tonnes, should again pass the 100 million tonne mark. Crude steel production should rise by a further 1.5 per cent in 1998. Capacity utilisation rates should rise to 67 per cent and 68 per cent in 1997 and 1998 respectively.

In 1996, steel imports were down by 15.4 per cent and the share of imports in the Japanese market rose to 8.8 per cent in 1995 from 7.5 per cent in 1996. At the same time, exports declined by 12.3 per cent, i.e. 2.7 million tonnes less than in 1995. In 1997, the increase in steel consumption will be more than matched by production and, as a result, net steel exports from Japan are expected to rise by almost 8 per cent. In contrast, net exports in 1998 will be slightly down by just under 4 per cent.

REPUBLIC OF KOREA

The Korean economy, which had grown at a rate of 9 per cent in 1994 and 1995, began to slow in 1996 when it grew by a mere 6.9 per cent. Export sales levelled off, partly as a result of the revaluation of the won, and imports of foreign goods increased. Employment continued to make progress, bringing a further fall in unemployment, down to below 2 per cent of the working population, and higher wages. Inflation rose slightly from 4.5 per cent in 1995 to 5 per cent in 1996. Industrial investment was up by 5 per cent and private consumption increased by 6.8 per cent. The Korean economy should follow a similar trend in 1997 with GDP growth of the order of 6 per cent. The balance of payments should improve, with the recovery in the world economy helping to fuel exports. Growth in investment should decline still further to little more than 3 per cent and private consumption should again increase by a further 6.5 per cent. The fiscal reforms undertaken by the government should lead to a gradual reduction in interest rates, which will nonetheless remain high in 1997. 1998 should see economic growth accelerate to 7 per cent.

Growth in activity in the construction sector dropped off in 1996, primarily due to the slowdown in the residential construction sector. The total number of new housing starts in 1997 should decline by 1.3 per cent, whereas government projects should help spur a 6.9 per cent increase in total investment in the construction sector. With regard to the automobile industry, domestic sales were up by 5.5 per cent in 1996, but growth will probably be limited to no more than 3.5 per cent in 1997. Car exports, which soared by 23.8 per cent in 1996, will probably grow by no more than 11.4 per cent in 1997. New orders in the shipbuilding industry were down by 2.8 per cent in 1996, and may well decline by a further 2 per cent or so in 1997.

Apparent steel consumption rose by 3.7 per cent, *i.e.* 1.3 million tonnes up on 1995, to 37 million tonnes of finished product equivalent. Crude steel production rose by 5.8 per cent, or 2.1 million tonnes, to 38.9 million tonnes. As a result of new steelmaking plant being brought into service, the capacity utilisation rate fell from 95 per cent in 1995 to 90 per cent in 1996. Steel imports rose by 3.2 per cent to pass the 10 million tonne mark. Exports were up by 11.4 per cent.

1997 should see a further increase in apparent steel consumption of around 2.8 per cent, amounting to slightly over 1 million tonnes. Crude steel production should increase by 5.7 per cent, *i.e.* an additional 2 million tonnes, taking the total to over 41 million tonnes. Exports should be slightly lower and imports down by 13 per cent compared with 1996.

Demand for steel should continue to grow at the same rate and there should be a further increase of almost 3 per cent. Steel exports should rise by 5.5 per cent, whereas imports should be slightly down. Crude steel production should grow by a further 5 per cent and, with an additional 2 million tonnes, should amount to over 43 million tonnes.

AUSTRALIA AND NEW ZEALAND

The **Australian** economy has grown steadily over the past five years against a background of low inflation. GDP grew by 4.1 per cent in 1996 and looks set to grow by 3.3 per cent in 1997 and by a further 3 per cent in 1998. Growth in the main steel-consuming sectors was lower in 1996, but output from the manufacturing and construction industries is expected to pick up in 1997. Investment in the mining sector is rising strongly and should help to maintain this trend.

As a result of these trends, apparent steel consumption, after four years of continuous growth, was down by 3.5 per cent. However, by as early as 1997 steel consumption should begin to rise again by 2.4 per cent and by a further 2.9 per cent in 1998. Crude steel production was down by just under 1 per cent in 1996, but this decline was primarily due to the renovation of a blast furnace at the Newcastle plant. 1997 should see steel production return to the 1995 level. In 1998, output should grow by around 1.5 per cent. Steel production capacity should gradually continue to rise and the utilisation rate should remain at a very high level of around 95 per cent.

Steel imports rose by 9.9 per cent in 1996 to 1.1 million tonnes. For the first time in many years Japan ceased to be Australia's main supplier and was replaced in this role by the European Union (15) whose exports to Australia doubled in 1996 compared with 1995. Exports were up by 8.1 per cent to a new record level of 3.2 million tonnes. Exports to the United States and Canada were up substantially, although Chinese Taipei remained the main market for Australian exports. Steel imports and exports are expected to be slightly down in 1997 and 1998.

The slowdown in the **New Zealand** economy resulted in a slight decline of 4.5 per cent in apparent steel consumption in 1996. By 1997 consumption is expected to have returned to the 1995 level and there may well be a further slight increase in 1998. A similar trend is anticipated in crude steel production, given that no significant changes are foreseen in crude steel production.

BRAZIL

Brazil has the largest economy in Latin America – accounting for almost half of the GDP of the area as a whole – and is fast becoming an increasingly important economic partner for neighbouring countries. Brazil is one of the keys to the economic stability of the region.

Brazil's current economic policy has been aimed at fully implementing the measures needed to consolidate the stabilisation of its currency and fiscal balance, on the one hand, and at restoring the conditions for sustainable development on the other. Brazil appears to have been successful in meeting these objectives in that, over the period 1993 to 1995, growth in GDP has averaged between 4.1 and 5.8 per cent and industrial output has grown at an average rate of around 5 per cent a year. However, in response to accelerating growth in GDP in early 1995, the government introduced restrictive measures which slowed the rate of growth during the second half of 1995 and early 1996.

Brazil's GDP should grow by around 3 per cent in 1996, but this rate of growth should increase to 4-5 per cent in 1997 and 1998. The monetary and fiscal policies introduced to counter inflation, which in the past had been running at levels of over 1 000 per cent, successfully brought it down to 14.8 per cent in 1995. Inflation should at most amount to 10 per cent in 1996 and should continue to fall in 1997 and 1998 to a level of 8 per cent. By the same token, the public deficit also seems to be under control. Other measures taken recently, such as the abolition of VAT on exports of semi-finished products, should allow production costs to be brought down. Lastly, the creation of Mercosur, a free-trade area for Latin America, should open up outlets for Brazilian industry and should be particularly beneficial to the steel industry.

Apparent steel consumption, expressed in finished product equivalent, would seem to have fallen by 0.6 per cent in 1996. This poor performance, after four years of continuous growth, is apparently due to a fall in consumption of long products as a result of stagnation in the construction sector. Net exports of steel products rose by 5.5 per cent, fuelling a 0.7 per cent increase in crude steel production.

1997 is expected to see higher demand for both flat and long products, primarily in response to high levels of activity in the automobile and construction sectors. Apparent steel consumption should strengthen to 6 per cent. Exports may well decline by 3 per cent, while crude steel production should be up by 2.1 per cent. In 1998, new investments arising from the current privatisation programmes, which are targeted on communications and infrastructure such as railways, ports and roads, should lead to a surge in demand for steel which might increase by as much as 14.6 per cent, or an additional 2 million tonnes. Crude steel production should rise by 4.3 per cent, while exports will continue to fall.

NON-OECD MARKET ECONOMIES

In 1996, these countries experienced contrasting economic trends. The rate of growth in the Dynamic Asian Economies was slower than in previous years, but still remained high at over 4.5 per cent. Growth in Latin American countries amounted to an average of 2.7 per cent, except for Venezuela which was in recession and whose economy contracted by 1 per cent. One of the reasons for the slowdown in economic activity may be seen in the restrictive economic policies adopted in order to bring down inflation. Once these policies have been successfully pursued, interest rates may start to come down, which will improve public deficits and encourage resumed growth at a pace that, although slower than in previous years, will probably be more sustainable. This should be the dominant trend in the years to come.

In 1996, total crude steel production in all non-OECD market economies, excluding Brazil, was 3.6 per cent up on the previous year. In South Africa, production fell by 8.8 per cent and in the rest of the African continent production fell for the fourth year in succession by a further 17.3 per cent. Production in Latin American, on the other hand, was 6.7 per cent up, *i.e.* an increase of 0.7 million tonnes. Production in Asia was up by 5 per cent and in the Middle East by 7.6 per cent.

1997 should see the decline in steel production in the African continent begin to slow to no more than 1.5 per cent in South Africa and 7.8 per cent in the rest of the continent. Crude steel production in Latin America should rise by 8 per cent, in the Middle East by 11 per cent and in Asia by 9.6 per cent. The increase in India will probably be no more than 3.2 per cent. 1998 is expected to bring higher steel production in all of these countries with the possible exception of a number of Asian countries (*inter alia* for technical reasons relating to the start-up of new capacity).

All of these countries, apart from South Africa, are net steel importers. Net steel imports for all of them was up by over 5 per cent in 1996, amounting to a total of 44.6 million tonnes. The largest increases were in net imports to the Middle East and the Dynamic Asian Economies. 1997 should see a significant reduction in net steel imports to these countries, down by 6.3 per cent, although such imports should start to resume an upward trend by as early as 1998, although at a lower rate than in previous years.

Apparent steel consumption continued to rise in these areas as a whole in 1996, although this increase had weakened to 4.8 per cent or some 5.2 million tonnes of finished product equivalent. Except in African countries, consumption rose in all areas at rates ranging from 4.5 to 6.7 per cent. 1997 should see a clear slowdown in steel consumption which might well decline in Latin America, South Africa and other African countries. The increase should amount to little more than 0.6 per cent in India and just over an average of 1 per cent in other Asian countries. It is only in the Middle East that steel demand might continue to rise steadily at a rate of over 6.5 per cent. 1998 should see a stronger recovery in steel demand.

The trend towards the expansion of existing steel production capacity was maintained in all of the countries within this area. In 1996, the capacity increase amounted to merely 4.1 per cent, *i.e.* some 4.5 million tonnes of additional capacity of which over 60 per cent in Asia and the rest in Latin America. The average capacity utilisation rate amounted to 69 per cent in 1996.

In 1997 almost 25 million tonnes of additional capacity should come on stream, an increase of 21.6 per cent over 1996. While the increase should remain modest in Latin America, at 1.7 million tonnes of additional capacity, and even more modest on the African continent, at 0.6 million additional tonnes, the increase in production capacity in the Middle East should amount to 2.6 million tonnes, that is 16.5 per cent. It is in Asia, however, that the increase in capacity will be the greatest at 20 million tonnes, an increase of 32.5 per cent. The average capacity utilisation rate for all areas combined should fall to no more than 60 per cent. In 1998 a further 33.7 million tonnes of additional steel production capacity should come on stream.

CENTRAL AND EASTERN EUROPE

Since the accession of Hungary, Poland and the Czech Republic to the OECD, this area now consists solely of Albania, Bulgaria, the Slovak Republic and Romania. The steel market in Albania may be treated as flat. Crude steel production in 1996 amounted to around 0.02 million tonnes and no major changes are anticipated over the next two years.

In **Bulgaria**, after two years of economic recovery, the country plunged back into deep recession. National output probably declined by 9 per cent. This led to a sharp devaluation of the currency and rampant inflation that reached 250 per cent. The situation should remain difficult in 1997 when GDP will probably fall by a further 3 per cent, and it will only be in 1998 that Bulgaria should start to emerge from the recession.

As a result of these economic difficulties, apparent steel consumption fell by 47.2 per cent in 1996 to a historic low of 0.5 million tonnes. In 1997, consumption should be up by 32 per cent, and by a further 36 per cent in 1998 when it should more or less regain the 1995 level. Steel imports plummeted by 67 per cent in 1996 to no more than 0.15 million tonnes, whereas exports were only down by 5.5 per cent at 1.55 million tonnes. Imports should gradually recover in 1997 but should then double in 1998, while exports should remain at a level of between 1.5 to 1.6 million tonnes over the next two years. Crude steel production fell by 9.6 per cent in 1996 to 2.5 million tonnes. It should remain stagnant at this level in 1997 before rising by over 4 per cent in 1998.

The economy of the **Slovak Republic** again expanded in 1996, growing by 6.9 per cent. Inflation remains under control at 5.3 per cent. Industrial output rose by 2.5 per cent, but the strongest growth proved to be in the building sector which grew by 4.4 per cent. Unemployment has been brought down to 12.6 per cent. GDP should continue to grow by a further 5.7 per cent in 1997 and probably by over

4 per cent in 1998. Inflation should remain under control at 5 to 6 per cent, while unemployment should continue on a downward trend to reach 12 per cent. Industrial output should rise by 4 per cent in 1997 and activity in the construction sector should increase by 3.5 per cent.

Apparent steel consumption in the Slovak Republic surged by 56 per cent in 1996, rising from 0.6 million to 0.9 million tonnes. Steel consumption is expected to remain at this level throughout 1997 and 1998. Crude steel production, on the other hand, fell by 9.4 per cent to no more than 3.6 million tonnes. Production should rise by 2 per cent in 1997 and by a further 5 per cent in 1998. Steel imports, on which there are no restrictions, grew by 4.2 per cent in 1996, 83 per cent of these imports coming from the Czech Republic. Imports should continue to grow by slightly over 6 per cent a year in 1997 and 1998. Exports declined by 16.3 per cent in 1996, falling from 3.4 to 2.9 million tonnes. However, they should climb back to 3 million tonnes by as early as 1997.

In **Romania**, the exchange rate crisis and the devaluation of the national currency, the lev, which fell 36 per cent against the dollar in 1996, prompted the government to introduce restrictions. Inflation soared to 59 per cent. GDP has nonetheless grown by 4.1 per cent and the rate of unemployment has fallen to 6.3 per cent, its lowest level in four years. In 1997, inflation may start to accelerate, rising to somewhere between 90 and 150 per cent, but after a sharp increase at the beginning of the year should start to subside in the second half of the year. Interest rates could rise substantially, too, and GDP may fall by 2 per cent. Conditions should return to normal, however, in 1998.

Apparent steel consumption was up by 11.4 per cent in 1996, thus rising back over the 3 million tonnes mark. A similar increase looks likely in 1997 and demand for steel should be up by a further 1.7 per cent in 1998. Crude steel production was down 7.3 per cent in 1996, but should recover with an increase of around 1.8 per cent in 1997 and almost 6 per cent in 1998 to take the total back over the 7 million tonnes mark. Steel imports, which had plummeted by 20 per cent in 1996, should start rising again in 1997 and 1998, while imports should remain relatively stable at a level of 0.5 to 0.6 million tonnes.

NEW INDEPENDENT STATES

In 1996, economic conditions remained difficult in most of the New Independent States (NIS) and, as in previous years, the situation varied substantially from one country to another. The GDP of the Russian Federation continued to decline, contracting by around 6 per cent. Industrial output fell yet again by a further 5 per cent and the volume of investment was down by 30 per cent. The rate of inflation remained high at around 147.7 per cent. It would seem that the prospects for recovery in 1997 will depend upon the ability of the Russian government to create the necessary conditions for an inflow of private capital into the economy. Domestic demand should already be strengthening. With regard to Ukraine, the decline in GDP slowed down in 1996 to a mere 7 per cent. Inflation also appears to be down, falling from 181 per cent in 1995 to no more than 50 per cent in 1996. The goals for structural and macroeconomic performance agreed upon jointly with the IMF appear to have been met. A new reform of the national currency has been successfully carried out and the central bank is preparing to make the currency convertible. All the indications are that the Ukrainian economy is set to start growing.

Apparent steel consumption in the NIS as a whole continued to decline and fell by 9.6 per cent in 1996 to 30 million tonnes of finished product equivalent. It should be noted that consumption in 1988 amounted to 117.8 million tonnes. This collapse in demand for steel is primarily due to weak domestic demand by end users and also to the critical financial situation of many firms. This decline in overall domestic demand for the NIS has led to a slight decrease in crude steel production, which was down by 1.1 per cent, but above all to a 12.5 per cent increase in exports which appear to have amounted to over 30 million tonnes in 1996. Of the main steel-producing countries, Russia is the only one whose production fell in 1996, by 4.2 per cent, to a level of 49.2 million tonnes, whereas production in both Ukraine and Kazakhstan rose by 4.6 and 3.8 per cent respectively.

In 1997, steel consumption in the NIS is expected to grow by 18.9 per cent, an increase of 5.7 million tonnes. This renewed growth should be accompanied not only by an increase of around 3 per cent in crude steel production, that is to say 2.3 million tonnes more than in 1996, but above all by a decline of almost

Capacities and steel production 1996

	Capacity	Production	Utilisation rate %
	Thousand tonnes		
Belarus	1 200	886	73.8
Kazakhstan	4 000	3 142	78.6
Federation of Russia	74 700	49 161	65.8
Ukraine	55 800	23 325	41.8
Uzbekistan	2 000	444	22.2
Total CIS	137 700	76 958	55.9
Azerbaïjan	800	18	2.3
Moldova	1 000	44	4.4
Georgia	1 800	646	35.9
Total NEI	141 300	77 666	55.0

Source: OECD, *Steelmaking Capacity in Non-OECD Countries.* Two-yearly Report.

4 million tonnes in exports. If consumption does start to increase, the trend should continue into 1998 and should again be accompanied by higher production and reduced exports.

CHINA AND NORTH KOREA

Although the data available on North Korea are both incomplete and unreliable, it is estimated that the country produced approximately 6 million tonnes of crude steel in 1996. This estimate has therefore been extended to both 1997 and 1998.

In China, inflation began to ease substantially from the middle of 1996 onwards and has not had any adverse impact on growth. After three years of tight credit controls, the annual rate of inflation had been brought down to 5.5 per cent by the early autumn. China's trade balance appears to have worsened. Growth in GDP remained at a high level of around 9.5 per cent in 1996 and should move even higher in 1997 to 9.8 per cent and possibly even 10 per cent in 1998. Domestic demand should follow a similar trend and should continue to climb in 1997 and 1998, as it did in 1996.

After successive declines in 1994 and 1995, apparent steel consumption rose by 11.2 per cent in 1996. This increase amounts to almost 11 million tonnes. Crude steel production was up by 5 million tonnes, an increase of 5.2 per cent, and the total output of 100.4 million tonnes in 1996 made China the world's largest producer of steel. Since stock surpluses were largely drawn down in 1995, exports from China of steel products fell by 27.9 per cent in 1996 to 6.9 million tonnes. In response to strengthening domestic demand, steel imports appear to have risen by around 18 per cent.

Domestic demand for steel in 1997 should continue to grow strongly by 4.8 per cent. Crude steel production should rise a further 1.4 per cent to some 101.7 million tonnes. Because of this, exports should fall off slightly while imports should continue to rise by some 12 per cent. Domestic steel demand may be slightly down by 0.7 per cent in 1998, but there should be continued growth of 1.2 per cent in crude steel production. Renewed demand on most world markets should lead to lower imports and higher exports.

4

TRENDS IN EMPLOYMENT IN THE STEEL INDUSTRY IN OECD MEMBER COUNTRIES

In 1996, the number of people employed in the steel industry in the OECD area fell by 28 400, a decline of around 3 per cent. Since 1974, the total number of jobs in the steel industry in the OECD area has fallen by 58 per cent.

In the European Union, the steel industry will continue to face growing international competition and the constant need to improve its competitiveness will again call for restructuring and downsizing. In the EU (15) as a whole, the workforce was reduced by around 6.8 per cent in 1996 with the loss of 22 500 jobs, the largest cuts being in Germany and Italy.

In other European countries, while the employment situation seems to have remained stable in Norway, the workforce in the Swiss steel industry has been cut by 7.7 per cent. In Turkey, in contrast, employment appears to have risen by 10.7 per cent.

Employment in the Japanese steel industry fell by 8.1 per cent, a loss of some 13 700 jobs. In Korea, on the other hand, employment in the steel industry grew by 1.8 per cent. Employment in the Mexican steel industry rose by a substantial 11.3 per cent.

Employment in the steel industry in Canada remained practically unchanged in 1996 and in 1997 should remain at a level of around 33 000 units, given that most of the restructuring of the industry has now been completed and that all plants are now operating at high capacity utilisation rates.

Despite higher output in the United States, employment in the steel industry has again declined by a further 1.3 per cent to an average level of 237 500 units. The cuts were proportionately lower in the production worker category, where the cuts amounted to a decrease of only 0.8 per cent.

STATISTICAL ANNEX

Table 1. **Apparent steel consumption** (million tonnes product equivalent)

Tableau 1. **Consommation apparente d'acier** (millions de tonnes équivalent produits)

	1990	1992	1993	1994	1995	1996	1997	1998	96/95 in/en %	97/96 in/en %	98/97 in/en %	
United States	86.0	82.7	89.7	101.6	97.5	103.5	100.4	101.5	+6.2	-3.0	+1.1	États-Unis
Canada	9.3	9.4	10.8	12.7	13.1	12.7	12.7	13.0	-2.8	+0.4	+1.7	Canada
EU (15)	119.2	115.1	102.4	113.6	127.0	111.0	116.6	123.2	-12.6	+5.0	+5.7	UE (15)
Other Europe [1]	25.0	19.2	21.1	22.1	28.8	24.1	25.5	28.0	-16.3	+5.6	+9.9	Autres Europe [1]
Japan	92.9	79.2	76.1	74.9	79.6	78.7	79.1	81.1	-1.2	+0.6	+2.5	Japon
Australia and New Zealand	5.3	5.2	5.6	6.0	6.3	6.1	6.3	6.4	-3.6	+2.5	+2.9	Australie et Nouvelle-Zélande
Mexico	6.6	8.2	8.6	10.1	8.3	8.8	9.8	10.8	+6.3	+10.6	+10.0	Mexique
Korea	19.6	21.6	25.1	30.3	35.7	37.0	38.1	39.2	+3.7	+2.8	+2.9	Corée
OECD	363.8	340.5	339.4	371.3	396.3	381.9	388.5	403.1	-3.6	+1.7	+3.8	OCDE
Brazil	8.6	8.7	9.3	11.0	12.1	12.0	12.7	14.6	-0.6	+6.0	+14.6	Brésil
Other Latin America	6.7	9.2	8.8	9.6	10.1	10.8	10.2	11.8	+6.7	-5.5	+16.0	Autres Amérique latine
South Africa	4.7	4.1	3.8	4.1	4.3	4.4	4.4	4.3	+2.8	-1.4	-0.5	Afrique du Sud
Other Africa	4.8	5.4	4.8	4.7	3.9	3.7	3.6	4.2	-5.2	-3.3	+18.3	Autres Afrique
Middle East	17.0	20.4	22.5	23.2	24.2	25.4	27.0	28.4	+4.6	+6.6	+4.9	Moyen-Orient
India	12.9	15.1	15.1	16.8	18.1	18.9	19.0	20.2	+4.5	+0.6	+6.2	Inde
Other Asia	33.3	44.3	40.8	42.9	48.5	51.1	51.7	51.4	+5.4	+1.2	-0.6	Autres Asie
Total	79.4	98.5	95.8	101.3	109.1	114.3	115.9	120.3	+4.8	+1.4	+3.8	Total
Total market economies	451.8	447.7	444.5	483.6	517.5	508.2	517.1	538.0	-1.8	+1.8	+4.0	Total économies de marché
Central and Eastern Europe	13.4	5.7	3.4	4.1	4.3	4.5	5.0	5.3	+5.6	+1.1	+5.8	Europe centrale et orientale
Romania	6.3	3.6	2.1	2.6	2.8	3.1	3.5	3.6	+11.8	+11.8	+1.7	Roumanie
Slovak Republic	5.7	1.4	0.7	0.6	0.6	0.9	0.9	0.9	+55.9	-2.2	+1.1	République slovaque
NIS	109.2	82.4	61.4	36.8	33.2	30.0	35.7	38.8	-9.6	+18.9	+8.7	NEI
China and North Korea	60.0	73.8	112.5	102.8	96.2	107.0	112.2	111.5	+11.2	+4.9	-0.6	Chine et Corée du Nord
World	634.3	609.6	621.7	627.2	651.1	649.6	669.8	693.5	-0.2	+3.1	+3.5	Monde

1. Includes: Norway, Turkey, Iceland, former Yougoslavia, the Czech Republic, Hungary and Poland.
Source: OECD Secretariat.

1. Comprend: Islande, Norvège, Turquie, ex-Yougoslavie, République tchèque, Hongrie et Pologne.
Source : Secrétariat de l'OCDE.

Table 2. **Trade balance (– = net exports, + = net imports)** (million tonnes)

Tableau 2. **Balance des échanges (– = exp. nettes, + = imp. nettes)** (millions de tonnes)

	1990	1991	1992	1993	1994	1995	1996	1997	1998	
EU (15)	-13.0	-16.1	-13.5	-26.8	-22.3	-12.4	-20.9	-18.3	-16.8	UE (15)
Japan	-9.5	-8.9	-12.1	-16.6	-16.6	-15.0	-13.3	-14.4	-13.8	Japon
Total	-22.5	-25.0	-26.4	-42.4	-38.9	-27.4	-34.2	-32.7	-30.6	Total
United States	11.7	8.5	11.6	14.1	23.8	15.7	21.9	16.4	14.2	États-Unis
Canada	-1.0	-2.1	-2.4	-1.5	0.8	0.7	0.1	-0.7	-0.8	Canada
Korea	-1.9	0.5	-4.6	-5.7	-1.2	1.4	0.7	-0.3	-1.1	Corée
Other Europe	-7.3	-7.9	-6.7	-5.3	-6.2	-0.5	-5.0	-4.8	-5.0	Autres Europe
Australia and New Zealand	-1.2	-1.3	-1.6	-2.2	-2.4	-2.1	-2.2	-2.2	-2.1	Australie et Nouvelle-Zélande
Mexico	-0.5	1.0	1.3	1.1	1.7	-1.7	-2.1	-1.8	-1.2	Mexique
Total	-0.2	-1.3	-2.4	0.5	16.5	13.5	13.4	6.6	4.0	Total
OECD	-22.7	-26.2	-28.0	-42.8	-22.3	-13.9	-20.8	-26.1	-26.6	OCDE
Brazil	-8.8	-10.7	-11.6	-12.0	-10.9	-9.4	-9.9	-9.6	-8.7	Brésil
Other Latin America	-1.0	0.2	1.3	0.7	0.8	1.0	1.0	-0.4	0.4	Autres Amérique latine
South Africa	-2.8	-3.4	-3.8	-3.9	-3.5	-3.5	-2.7	-2.7	-3.8	Afrique du Sud
Other Africa	3.2	2.8	3.8	3.6	3.5	2.8	2.8	2.7	3.2	Autres Afrique
Middle East	11.2	12.0	12.5	13.0	13.1	13.8	14.1	14.5	15.5	Moyen-Orient
India	1.1	0.5	0.7	0.1	0.8	0.9	0.7	0.2	-0.1	Inde
Other Asia	18.2	24.0	27.5	21.4	24.0	27.3	29.0	27.5	28.2	Autres Asie
Total	29.9	36.1	42.0	34.9	38.7	42.3	46.0	41.8	43.4	Total
Total market economies	-1.6	-0.8	2.4	-19.9	5.5	19.0	15.3	6.1	8.1	Total économies de marché
Central and Eastern Europe	-0.3	-0.2	-2.9	-5.6	-5.5	-6.3	-5.2	-5.4	-5.7	Europe centrale et orientale
Romania	-1.3	-0.8	-0.6	-2.2	-2.0	-2.4	-1.7	-1.9	-2.2	Roumanie
Slovak Republic	1.3	0.9	-1.7	-2.5	-2.6	-2.7	-2.1	-2.2	-2.4	République slovaque
NIS	-1.4	-0.7	-2.1	-9.7	-20.6	-25.1	-28.0	-24.2	-23.5	NEI
China and North Korea	1.6	0.0	2.9	34.1	21.1	13.1	19.8	23.5	21.0	Chine et Corée du Nord
Unspecified	1.8	1.8	0.3	1.0	0.6	0.6	0.7	0.1	0.0	Non spécifié

Source: OECD Secretariat.

Source : Secrétariat de l'OCDE.

Table 3. **Crude steel production** (million tonnes)

Tableau 3. **Production d'acier brut** (millions de tonnes)

	1990	1992	1993	1994	1995	1996	1997	1998	96/95 in/en %	97/96 in/en %	98/97 in/en %	
United States	89.7	84.3	88.8	91.2	95.2	94.7	97.5	101.3	-0.5	+3.0	+3.9	États-Unis
Canada	12.3	13.9	14.4	13.9	14.4	14.6	15.6	16.0	+1.5	+6.6	+2.6	Canada
EU (15)	148.4	143.8	144.3	151.7	155.6	147.0	150.3	155.9	-5.5	+2.3	+3.7	UE (15)
Other Europe	40.7	32.1	32.5	34.7	35.7	34.7	35.8	38.8	-2.9	+3.2	+8.5	Autres Europe
Japan	110.3	98.1	99.6	98.3	101.6	98.8	100.4	102.0	-2.8	+1.7	+1.5	Japon
Australia and New Zealand	7.4	7.7	8.6	9.2	9.3	9.2	9.3	9.5	-1.0	+1.5	+1.5	Australie et Nouvelle-Zélande
Korea	23.1	28.1	33.0	33.8	36.8	38.9	41.1	43.2	+5.8	+5.7	+5.0	Corée
Mexico	8.7	8.5	9.2	10.3	12.2	13.2	13.9	14.3	+8.4	+5.5	+3.1	Mexique
OECD	440.7	416.4	430.4	443.0	460.7	451.0	463.9	480.9	-2.1	+2.9	+3.7	OCDE
Brazil	20.6	23.9	25.2	25.7	25.1	25.3	25.8	26.9	+0.7	+2.1	+4.3	Brésil
Other Latin America	9.2	9.1	9.3	10.1	10.5	11.2	12.2	13.1	+6.7	+8.2	+7.6	Autres Amérique latine
South Africa	8.6	9.1	8.7	8.5	8.7	8.0	7.9	9.1	-8.8	-1.5	+15.4	Afrique du Sud
Other Africa	2.0	2.1	1.6	1.5	1.4	1.2	1.1	1.2	-17.3	-7.8	+17.0	Autres Afrique
Middle East	6.3	8.9	10.6	11.3	11.7	12.6	13.9	14.3	+7.6	+11.0	+2.6	Moyen-Orient
India	15.0	18.1	18.2	19.3	20.8	21.8	22.5	24.2	+5.0	+3.2	+7.6	Inde
Other Asia	16.7	18.5	21.2	20.6	23.1	24.3	26.6	25.4	+5.0	+9.6	-4.5	Autres Asie
Total	57.8	65.8	69.6	71.3	76.2	79.1	84.2	87.3	+3.6	+6.5	+3.8	Total
Total market economies	519.1	506.1	525.2	540.0	562.0	555.4	573.9	595.1	-1.2	+3.3	+3.7	Total économies de marché
Central and Eastern Europe	17.4	10.8	11.3	12.2	13.2	12.1	12.9	13.6	-8.4	+6.5	+5.3	Europe centrale et orientale
Romania	9.8	5.4	5.5	5.8	6.6	6.1	6.8	7.2	-7.3	+1.8	+5.9	Roumanie
Slovak Republic	5.5	3.9	3.9	4.0	3.9	3.6	3.6	3.8	-9.4	+2.0	+5.0	République slovaque
NIS	154.4	118.0	98.1	78.3	78.8	78.0	80.3	83.5	-1.1	+3.0	+4.0	NEI
China and North Korea	73.4	87.9	96.5	99.6	101.4	106.4	107.8	109.0	+4.9	+1.3	+1.2	Chine et Corée du Nord
World	764.1	722.7	731.0	730.0	755.4	751.7	774.8	801.2	-0.5	+3.1	+3.4	Monde

Source: OECD Secretariat.

Source : Secrétariat de l'OCDE.

Table 4. **Steel production, consumption and trade** (million tonnes)

Tableau 4. **Production, consommation et échanges d'acier** (millions de tonnes)

1995	Production			Imports/ Importations	Exports/ Exportations	Balance	Apparent consumption/ Consommation apparente	1995
	Crude steel/ Acier brut	via/c.c.	Product eq./ Équiv. produits					
United States	95.2	86.6	81.7	22.1	6.4	15.7	97.5	États-Unis
Canada	14.4	14.0	12.4	5.2	4.6	0.7	13.1	Canada
EU (15)	155.6	145.1	139.4	20.1	32.5	−12.4	127.0	UE (15)
Other Europe	35.7	20.6	29.3	13.8	13.4	−0.5	28.8	Autres Europe
Japan	101.6	97.4	94.6	7.0	22.0	−15.0	79.6	Japon
Australia and New Zealand	9.3	9.3	8.4	1.3	3.4	−2.1	6.3	Australie et Nouvelle-Zélande
Korea	36.8	35.9	34.3	9.8	8.4	1.4	35.7	Corée
Mexico	12.2	9.1	10.0	0.7	2.4	−1.7	8.3	Mexique
OECD	460.7	418.1	410.2	79.5	93.4	−13.9	396.3	OCDE
Brazil	25.1	15.9	21.4	0.3	9.7	−9.4	12.1	Brésil
Other Latin America	10.5	8.8	9.1	5.3	4.4	1.0	10.1	Autres Amérique latine
South Africa	8.7	7.9	7.8	0.3	3.8	−3.5	4.3	Afrique du Sud
Other Africa	1.4	0.4	1.1	3.2	0.4	2.8	3.9	Autres Afrique
Middle East	11.7	11.3	10.5	14.8	1.1	13.8	24.2	Moyen-Orient
India	20.8	10.9	17.2	2.2	1.3	0.9	18.1	Inde
Other Asia	23.1	22.6	21.2	34.9	7.6	27.3	48.5	Autres Asie
Total	76.2	61.9	66.9	60.7	18.6	42.3	109.1	Total
Total market economies	562.0	495.9	498.5	140.5	121.7	19.0	517.5	Total économies de marché
Central and Eastern Europe	13.2	7.3	10.6	1.6	7.9	−6.3	4.3	Europe centrale et orientale
Romania	6.6	3.1	5.2	0.4	2.8	−2.4	2.8	Roumanie
Slovak Republic	3.9	3.5	3.3	0.7	3.4	−2.7	0.6	République slovaque
NIS	78.8	29.3	58.3	2.2	27.3	−25.1	33.2	NEI
China and North Korea	101.4	38.1	83.1	22.6	9.5	13.1	96.2	Chine et Corée du Nord
World	755.4	568.4	650.4	166.8	166.2	0.6	651.1	Monde

Source: OECD Secretariat.

Source : Secrétariat de l'OCDE.

Table 5. **Steel production, consumption and trade** (million tonnes)

Tableau 5. **Production, consommation et échanges d'acier** (millions de tonnes)

1996	Production			Imports/ Importations	Exports/ Exportations	Balance	Apparent consumption/ Consommation apparente	1996
	Crude steel/ Acier brut	via/c.c.	Product eq./ Équiv. produits					
United States	94.7	88.3	81.6	26.5	4.6	21.9	103.5	États-Unis
Canada	14.6	14.3	12.6	4.5	4.4	0.1	12.7	Canada
EU (15)	147.0	138.4	131.9	13.0	33.9	−20.9	111.0	UE (15)
Other Europe	34.7	24.4	29.1	11.2	16.2	−5.0	24.1	Autres Europe
Japan	98.8	95.2	92.0	5.9	19.3	−13.3	78.7	Japon
Australia and New Zealand	9.2	9.2	8.3	1.4	3.6	−2.2	6.1	Australie et Nouvelle-Zélande
Korea	38.9	38.0	36.3	10.1	9.4	0.7	37.0	Corée
Mexico	13.2	10.5	11.0	0.7	2.9	−2.1	8.8	Mexique
OECD	451.0	418.3	402.7	73.3	94.2	−20.9	381.9	OCDE
Brazil	25.3	18.1	21.9	0.4	10.3	−9.9	12.0	Brésil
Other Latin America	11.2	9.8	9.8	4.5	3.5	1.0	10.8	Autres Amérique latine
South Africa	8.0	7.2	7.1	0.3	3.0	−2.7	4.4	Afrique du Sud
Other Africa	1.2	0.3	0.9	3.1	0.4	2.8	3.7	Autres Afrique
Middle East	12.6	12.1	11.3	15.0	1.0	14.1	25.4	Moyen-Orient
India	21.8	10.9	18.2	2.4	1.8	0.7	18.9	Inde
Other Asia	24.3	22.2	22.1	35.7	6.7	29.0	51.1	Autres Asie
Total	79.1	62.5	69.4	61.0	16.4	44.6	114.3	Total
Total market economies	555.4	498.9	494.0	134.7	120.9	13.8	508.2	Total économies de marché
Central and Eastern Europe	12.1	7.2	9.8	1.5	6.7	−5.2	4.5	Europe centrale et orientale
Romania	6.1	3.1	4.8	0.6	2.3	−1.7	3.1	Roumanie
Slovak Republic	3.6	3.6	3.0	0.8	2.9	−2.1	0.9	République slovaque
NIS	78.0	31.7	58.0	2.7	30.7	−28.0	30.0	NEI
China and North Korea	106.4	40.0	87.2	26.7	6.9	19.8	107.0	Chine et Corée du Nord
World	751.7	577.8	648.9	165.6	164.9	0.7	649.6	Monde

Source: OECD Secretariat.

Source : Secrétariat de l'OCDE.

Table 6. **Steel production, consumption and trade** (million tonnes)

Tableau 6. **Production, consommation et échanges d'acier** (millions de tonnes)

1996	Production			Imports/ Importations	Exports/ Exportations	Balance	Apparent consumption/ Consommation apparente	1996
	Crude steel Acier brut	via/c.c.	Product eq./ Équiv. produits					
United States	97.5	90.9	84.0	21.7	5.3	16.4	100.4	États-Unis
Canada	15.6	15.2	13.4	3.5	4.2	-0.7	12.7	Canada
EU(15)	150.3	141.6	134.9	13.2	31.4	-18.3	116.6	UE(15)
Other Europe	35.8	26.4	30.3	11.6	16.4	-4.8	25.5	Autres Europe
Japan	100.4	96.8	93.5	5.1	19.5	-14.4	79.1	Japon
Australia and New Zealand	9.3	9.3	8.4	1.5	3.6	-2.2	6.3	Australie et Nouvelle-Zélande
Korea	41.1	40.2	38.4	8.8	9.1	-0.3	38.1	Corée
Mexico	13.9	11.1	11.6	0.8	2.6	-1.8	9.8	Mexique
OECD	463.9	431.6	414.5	66.2	92.2	-26.0	388.5	OCDE
Brazil	25.8	18.5	22.3	0.3	10.0	-9.6	12.7	Brésil
Other Latin America	12.2	10.6	10.6	4.3	4.7	-0.4	10.2	Autres Amérique latine
South Africa	7.9	7.1	7.0	0.4	3.0	-2.7	4.4	Afrique du Sud
Other Africa	1.1	0.3	0.9	3.0	0.3	2.7	3.6	Autres Afrique
Middle East	13.9	13.5	12.5	15.5	1.0	14.5	27.0	Moyen-Orient
India	22.5	11.3	18.8	2.2	2.0	0.2	19.0	Inde
Other Asia	26.6	24.0	24.2	35.1	7.6	27.5	51.7	Autres Asie
Total	84.2	66.8	74.0	60.5	18.6	41.8	115.9	Total
Total market economies	573.9	516.9	510.8	127.0	120.8	6.2	517.1	Total économies de marché
Central and Eastern Europe	12.9	7.6	10.4	1.6	6.9	-5.4	5.0	Europe centrale et orientale
Romania	6.8	3.4	5.4	0.5	2.4	-1.9	3.5	Roumanie
Slovak Republic	3.6	3.6	3.1	0.8	3.0	-2.2	0.9	République slovaque
NIS	80.3	33.7	59.9	2.3	26.5	-24.2	35.7	NEI
China and North Korea	107.8	43.1	88.7	30.0	6.5	23.5	112.2	Chine et Corée du Nord
World	774.8	601.3	669.7	160.7	160.5	0.1	669.8	Monde

Source: OECD Secretariat.

Source : Secrétariat de l'OCDE.

Table 7. **Steel production, consumption and trade** (million tonnes)

Tableau 7. **Production, consommation et échanges d'acier** (millions de tonnes)

1998	Production			Imports/ Importations	Exports/ Exportations	Balance	Apparent consumption/ Consommation apparente	1998
	Crude steel/ Acier brut	via/c.c.	Product eq./ Équiv. produits					
United States	101.3	94.5	87.3	20.2	6.0	14.2	101.5	États-Unis
Canada	16.0	15.6	13.8	3.2	4.0	-0.8	13.0	Canada
EU (15)	155.9	147.2	140.0	12.0	28.7	-16.8	123.2	UE (15)
Other Europe	38.8	30.4	33.0	11.4	16.5	-5.0	28.0	Autres Europe
Japan	102.0	98.3	94.9	5.5	19.3	-13.8	81.1	Japon
Australia and New Zealand	9.5	9.5	8.5	1.2	3.3	-2.1	6.4	Australie et Nouvelle-Zélande
Korea	43.2	42.2	40.3	8.5	9.6	-1.1	39.2	Corée
Mexico	14.3	11.5	11.9	0.7	1.9	-1.2	10.8	Mexique
OECD	480.9	448.9	429.7	62.7	89.3	-26.6	403.1	OCDE
Brazil	26.9	19.3	23.3	0.4	9.1	-8.7	14.6	Brésil
Other Latin America	13.1	11.6	11.4	5.5	5.0	0.4	11.8	Autres Amérique latine
South Africa	9.1	8.2	8.1	0.3	4.0	-3.8	4.3	Afrique du Sud
Other Africa	1.2	0.4	1.0	3.5	0.3	3.2	4.2	Autres Afrique
Middle East	14.3	13.8	12.9	16.6	1.1	15.5	28.4	Moyen-Orient
India	24.2	12.1	20.2	2.0	2.1	-0.1	20.2	Inde
Other Asia	25.4	23.3	23.2	35.4	7.2	28.2	51.4	Autres Asie
Total	87.3	69.4	76.8	63.3	19.7	43.4	120.3	Total
Total market economies	595.1	537.6	529.8	126.4	118.1	8.3	538.0	Total économies de marché
Central and Eastern Europe	13.6	8.4	11.0	2.0	7.6	-5.7	5.3	Europe centrale et orientale
Romania	7.2	4.0	5.8	0.6	2.8	-2.2	3.6	Roumanie
Slovak Republic	3.8	3.8	3.3	0.9	3.2	-2.4	0.9	République slovaque
NIS	83.5	35.1	62.3	2.5	26.0	-23.5	38.8	NEI
China and North Korea	109.0	49.1	90.5	29.0	8.0	21.0	111.5	Chine et Corée du Nord
World	801.2	630.1	693.5	159.6	159.6	0.0	693.5	Monde

Source: OECD Secretariat.

Source : Secrétariat de l'OCDE.

Table 8. **The steel markets in the United States, the EU and Japan**

Tableau 8. **Les marchés de l'acier aux États-Unis, dans l'UE et au Japon**

	United States/États-Unis				EU (15)/UE (15)				Japan/Japon				
	1995	1996	1997	1998	1995	1996	1997	1998	1995	1996	1997	1998	
	In million product tonnes/Millions de tonnes produit												
Real consumption	98.2	101.5	101.4	101.9	121.0	119.0	116.5	122.6	79.2	78.9	78.9	80.8	Consommation réelle
Stocks, consumers and merchants	–0.6	+0.8	–0.2	–0.2	+3.0	–4.0	+0.2	+0.5	+0.4	–0.2	+0.2	+0.3	Stocks des consommateurs et des marchands
Market	97.6	102.3	101.2	101.7	124.0	115.0	116.7	123.1	79.6	78.7	79.1	81.1	Marché
Imports	22.1	26.5	21.7	20.2	20.1	13.0	13.2	12.0	7.0	5.9	5.1	5.5	Importations
Exports	6.4	4.6	5.3	6.0	32.5	33.9	31.4	28.7	22.0	19.3	11.5	19.3	Exportations
Deliveries	81.9	80.4	84.8	87.5	136.4	135.9	134.9	139.8	94.6	92.0	93.5	94.9	Livraisons
Stocks producers	–0.2	+1.2	–0.8	–0.2	+3.0	–4.0	0.0	+0.2	0	0	0	0	Stocks des producteurs
Production	81.7	81.6	84.0	87.3	139.4	131.9	134.9	140.0	94.6	92.0	93.5	94.9	Production
	In million tonnes of crude steel/Millions tonnes d'acier brut												
Crude steel production	95.2	94.7	97.5	101.3	155.6	147.0	150.3	155.9	101.6	98.8	100.4	102.0	Production d'acier brut
Capacity	102.1	105.2	109.3	110.2	201.3	198.8	200.3	202.2	138.0	149.7	149.7	149.7	Capacité
	In %/En %												
Capacity utilisation	93.2	90.0	89.2	91.9	77.3	73.9	75.0	77.1	73.6	66.0	67.1	68.1	Utilisation de capacité
Import share	22.6	25.9	21.4	19.9	16.2	11.3	11.3	9.7	8.8	7.5	6.4	6.8	Part d'importation

Source: OECD Secretariat.

Source : Secrétariat de l'OCDE.

Table 9. **Steel markets in EU countries** (in million tonnes product equivalent)

Tableau 9. **Les marchés de l'acier dans les pays de l'UE** (en millions de tonnes équivalent produits)

	Germany/Allemagne				France				Italy/Italie				United Kingdom/Royaume-Uni				
	1995	1996	1997	1998	1995	1996	1997	1998	1995	1996	1997	1998	1995	1996	1997	1998	
Real consumption	35.6	31.1	35.2	36.6	14.2	14.8	14.8	15.0	22.5	23.2	19.9	21.9	13.3	13.2	14.0	14.4	Consommation réelle
Stocks	+0.9	-0.5	-0.1	-0.2	+1.0	-0.9	-0.2	+0.5	+1.5	-1.4	+0.3	0	+0.6	-0.1	+0.2	+0.1	Stocks
Apparent consumption	36.5	30.7	35.1	36.4	15.2	13.9	14.6	15.5	24.0	21.8	20.2	21.9	13.9	13.1	14.2	14.5	Consommation apparente
Imports	19.3	15.4	16.6	16.6	10.1	10.6	10.0	9.5	11.8	10.5	10.0	10.0	6.5	5.7	5.8	6.0	Importations
Exports	20.5	20.5	18.4	17.5	11.1	12.5	12.5	11.5	12.7	10.5	11.5	12.0	8.2	8.6	8.2	8.2	Exportations
Production	37.6	35.7	36.8	37.3	16.2	15.8	17.1	17.5	25.0	21.8	21.7	23.9	15.6	16.0	16.6	16.7	Production
Crude steel production	41.9	39.8	41.0	41.5	17.8	17.6	19.1	19.6	27.8	24.3	24.1	26.5	17.6	18.0	18.7	18.8	Production d'acier brut
Capacity	50.5	50.7	50.7	51.6	23.2	22.0	22.7	22.7	41.1	41.8	41.8	41.8	20.5	20.7	20.7	20.7	Capacité
Capacity utilisation in %	83	79	81	80	78	80	84	86	68	58	58	63	86	87	90	91	Utilisation de la capacité en %

	Netherlands/Pays-Bas				Belgium and Luxembourg/Belgique et Luxembourg				Spain/Espagne				Rest EU (15)/Reste UE (15)				
	1995	1996	1997	1998	1995	1996	1997	1998	1995	1996	1997	1998	1995	1996	1997	1998	
Real consumption	4.8	4.4	4.5	4.5	3.9	3.7	3.5	3.8	12.5	11.4	11.1	12.3	5.9	5.9	5.9	5.9	Consommation réelle
Stocks	+0.3	-0.4	0	+0.2	+0.6	-0.3	-0.2	+0.2	+0.6	-0.6	+0.4	+0.1	+0.3	-0.2	+0.2	-0.1	Stocks
Apparent consumption	5.1	4.0	4.5	4.7	4.5	3.4	3.3	4.0	13.1	10.8	11.5	12.4	6.2	5.7	5.7	5.8	Consommation apparente
Imports	5.7	4.9	5.1	5.1	5.3	6.1	5.5	5.0	4.8	4.5	4.5	4.7	5.9	5.7	5.8.	5.6	Importations
Exports	6.4	6.5	6.3	6.1	13.3	14.6	13.9	13.4	4.1	4.6	4.0	4.2	2.2	2.6	2.6	2.5	Exportations
Production	5.8	5.7	5.7	5.7	12.5	11.9	11.6	12.4	12.4	10.9	11.0	11.9	2.5	2.5	2.6	2.7	Production
Crude steel production	6.4	6.3	6.3	6.4	14.2	13.3	13.1	13.9	13.8	12.2	12.3	13.3	2.7	2.8	2.9	3.0	Production d'acier brut
Capacity	6.5	6.8	6.8	6.8	19.9	18.8	18.8	18.8	19.5	17.7	18.4	18.4	6.1	6.0	6.1	6.1	Capacité
Capacity utilisation in %	98	93	93	94	71	71	70	74	71	69	67	72	45	47	47	48	Utilisation de la capacité en %

	Austria/Autriche				Finland/Finlande				Sweden/Suède				
	1995	1996	1997	1998	1995	1996	1997	1998	1995	1996	1997	1998	
Real consumption	3.0	2.8	2.8	2.6	2.1	2.1	2.0	2.1	3.2	3.3	3.2	3.3	Consommation réelle
Stocks	+0.1	-0.2	-0.2	+0.1	-0.2	-0.1	0	0	+0.3	-0.3	-0.2	-0.2	Stocks
Apparent consumption	3.1	2.6	2.6	2.7	1.9	2.0	2.0	2.1	3.5	3.0	3.0	3.1	Consommation apparente
Imports	2.0	1.6	1.7	1.5	1.1	1.1	1.1	1.1	2.8	2.1	2.2	2.1	Importations
Exports	3.4	3.0	3.2	3.1	2.1	2.2	2.3	2.1	3.8	3.6	3.7	3.4	Exportations
Production	4.5	4.0	4.1	4.3	3.0	3.1	3.2	3.2	4.5	4.4	4.5	4.4	Production
Crude steel production	5.0	4.4	4.5	4.8	3.2	3.3	3.4	3.4	5.0	4.9	5.1	4.9	Production d'acier brut
Capacity	4.6	4.6	4.7	5.6	4.0	4.2	4.2	4.2	5.4	5.5	5.5	5.5	Capacité
Capacity utilisation in %	108	97	96	86	80	79	80	81	92	89	92	89	Utilisation de la capacité en %

Note. Trade figures for individual EU-countries represent the sum of intra-EU trade and trade with third countries.

Source. OECD Secretariat.

Note : Les chiffres d'échanges pour les pays individuels des UE représentent la somme des échanges avec les pays tiers et des échanges intra-communautaires.

Source : Secrétariat de l'OCDE.

Table 10. **Steel markets in "Other Western Europe"** (in million tonnes product equivalent)

Tableau 10. **Les marchés de l'acier dans les autres pays d'Europe occidentale et le Mexique** (en millions de tonnes équivalent produits)

	Turkey/Turquie				Iceland and ex-Yugoslavia/ Islande et ex-Yougoslavie				Norway/Norvège				Switzerland/Suisse			
	1995	1996	1997	1998	1995	1996	1997	1998	1995	1996	1997	1998	1995	1996	1997	1998
Apparent consumption — Consommation apparente	13.1	9.9	10.2	10.2	0.9	1.3	2.4	3.3	1.4	1.3	1.3	1.3	1.9	1.6	1.6	1.7
Imports — Importations	5.6	3.9	3.4	3.2	0.8	0.9	1.5	2.0	1.6	1.6	1.5	1.5	2.0	1.6	1.7	1.8
Exports — Exportations	3.9	6.0	6.3	6.5	0.5	0.5	0.6	0.6	0.7	0.7	0.7	0.6	0.8	0.7	0.8	0.9
Production — Production	11.4	12.0	13.0	13.5	0.6	1.0	1.5	1.9	0.5	0.5	0.5	0.5	0.7	0.8	0.7	0.8
Crude steel production — Production d'acier brut	12.8	13.4	14.6	15.1	0.7	1.1	1.7	2.1	0.5	0.5	0.5	0.5	0.8	0.9	0.7	0.9
Capacity — Capacité	14.8	19.3	19.5	19.8	2.5	2.5	2.5	2.8	0.6	0.6	0.6	0.6	1.1	1.1	1.1	1.1
Capacity utilisation in % — Utilisation de capacité en %	86	69	75	76	28	44	68	75	93	91	98	91	78	81	70	86

	Czech Rep./Rép. tchèque				Hungary/Hongrie				Poland/Pologne				Mexico/Mexique			
	1995	1996	1997	1998	1995	1996	1997	1998	1995	1996	1997	1998	1995	1996	1997	1998
Apparent consumption — Consommation apparente	3.5	3.0	3.0	3.7	1.4	1.2	1.3	1.3	6.6	5.8	5.8	6.6	8.3	8.8	9.8	10.8
Imports — Importations	1.5	1.6	1.6	1.4	0.7	0.6	0.6	0.6	1.0	1.2	1.3	1.0	0.7	0.7	0.8	0.7
Exports — Exportations	3.6	3.7	3.6	3.5	0.9	0.9	1.0	1.0	3.5	3.6	3.5	3.4	2.4	2.9	2.6	1.9
Production — Production	5.5	5.2	5.0	5.8	1.6	1.6	1.6	1.7	9.1	8.2	8.0	9.0	10.0	11.0	11.6	11.9
Crude steel production — Production d'acier brut	7.2	6.5	6.3	7.4	1.9	1.9	1.9	1.9	11.9	10.4	10.1	11.1	12.2	13.2	13.9	14.3
Capacity — Capacité	11.0	8.8	8.8	8.8	2.4	2.5	1.9	1.9	14.1	11.7	12.5	12.3	11.0	15.2	16.3	16.3
Capacity utilisation in % — Utilisation de capacité en %	65	74	72	84	78	75	97	100	84	89	81	90	111	86	85	88

Source: OECD Secretariat.

Source : Secrétariat de l'OCDE.

Table 11. **Manpower**

Tableau 11. **Main-d'œuvre**

	Average numbers employed ('000)/Moyenne des effectifs ('000)							% change/variation		
	1974	1984	1992	1993	1994	1995	1996	1996/95	1996/74	
Belgium/Luxembourg	86.6	51.4	34.0	32.2	30.9	29.8	28.4	-4.7	-67.2	Belgique/Luxembourg
Denmark, Ireland	3.5	2.3	1.9	1.8	1.6	1.5	1.5	–	-57.1	Danemark, Irlande
France	155.7	87.1	43.9	41.2	40.4	39.3	38.7	-1.5	-75.1	France
Germany	230.6	156.5	137.4	119.0	100.0	92.5	85.9	-7.1	-62.7	Allemagne
Greece	8.7	4.2	3.1	3.0	2.7	2.5	2.3	-8.0	-73.6	Grèce
Italy	93.8	81.7	52.0	50.4	45.5	42.1	39.3	-6.7	-58.1	Italie
Netherlands	23.8	18.7	16.3	14.6	13.1	12.6	12.3	-2.4	-48.3	Pays-Bas
Portugal	5.0	6.7	3.4	3.2	2.9	2.7	2.4	-11.1	-52.0	Portugal
Spain	89.4	69.2	34.7	30.1	26.7	25.3	23.8	-5.9	-73.3	Espagne
United Kingdom	197.7	62.3	42.4	40.2	38.5	37.9	37.2	-1.8	-81.2	Royaume-Uni
EU (12)	894.8	540.1	369.0	335.4	302.3	286.2	271.8	-5.0	-69.6	UE (12)
Austria	43.0	34.9	17.9	16.2	15.4	14.9	12.9	-13.4	-70.0	Autriche
Finland	8.1	9.0	8.5	8.7	8.8	9.0	9.0	–	+11.1	Finlande
Sweden	51.0	32.2	21.7	20.9	20.7	20.7	14.6	-29.5	-71.4	Suède
EU (15)	996.9	616.2	417.1	381.2	347.2	330.8	308.3	-6.8	-69.1	UE (15)
Norway	7.3[1]	4.0	1.4	1.4	1.3	1.3[1]	1.3[1]	–	-82.2	Norvège
Switzerland	5.2	3.0	2.2	1.9	1.6	1.3	1.2	-7.7	-76.9	Suisse
Turkey	36.1	35.0	36.8	35.2	32.4	29.9[1]	32.0[1]	+10.7	-11.4	Turquie
Canada	52.2	51.5	34.8	33.4	31.5	33.7	33.5	-0.6	-35.8	Canada
United States	609.5	267.4	253.5	238.8	233.5	240.7	237.5	-1.3	-61.0	États-Unis
Australia	43.2	30.5	26.3[1]	26.3[1]	26.0[1]	26.0[1]	26.0[1]	–	-39.8	Australie
Japan	323.9	264.8	189.6	193.0	182.7	168.8	155.1	-8.1	-52.1	Japon
Mexico	45.5	77.6[1]	44.8	56.8	57.0	48.6	55.0[1]	+11.3	+20.9	Mexique
Korea	62.9[1]	62.9	67.7	66.2	59.8[1]	66.3[1]	67.5[1]	+1.8	+5.4	Corée
Total OECD	2183.0[1]	1412.9[1]	1074.4[1]	1034.5	1013.4	945.9[1]	917.5[1]	-3.0	-58.0	Total OCDE

1. Secretariat estimate.

1. Estimation du Secrétariat.

MAIN SALES OUTLETS OF OECD PUBLICATIONS
PRINCIPAUX POINTS DE VENTE DES PUBLICATIONS DE L'OCDE

AUSTRALIA – AUSTRALIE
D.A. Information Services
648 Whitehorse Road, P.O.B 163
Mitcham, Victoria 3132   Tel. (03) 9210.7777
                                Fax: (03) 9210.7788

AUSTRIA – AUTRICHE
Gerold & Co.
Graben 31
Wien I   Tel. (0222) 533.50.14
                           Fax: (0222) 512.47.31.29

BELGIUM – BELGIQUE
Jean De Lannoy
Avenue du Roi, Koningslaan 202
B-1060 Bruxelles   Tel. (02) 538.51.69/538.08.41
                                Fax: (02) 538.08.41

CANADA
Renouf Publishing Company Ltd.
5369 Canotek Road
Unit 1
Ottawa, Ont. K1J 9J3   Tel. (613) 745.2665
                                Fax: (613) 745.7660

Stores:
71 1/2 Sparks Street
Ottawa, Ont. K1P 5R1   Tel. (613) 238.8985
                                Fax: (613) 238.6041

12 Adelaide Street West
Toronto, QN M5H 1L6   Tel. (416) 363.3171
                                Fax: (416) 363.5963

Les Éditions La Liberté Inc.
3020 Chemin Sainte-Foy
Sainte-Foy, PQ G1X 3V6   Tel. (418) 658.3763
                                Fax: (418) 658.3763

Federal Publications Inc.
165 University Avenue, Suite 701
Toronto, ON M5H 3B8   Tel. (416) 860.1611
                                Fax: (416) 860.1608

Les Publications Fédérales
1185 Université
Montréal, QC H3B 3A7   Tel. (514) 954.1633
                                Fax: (514) 954.1635

CHINA – CHINE
Book Dept., China National Publications
Import and Export Corporation (CNPIEC)
16 Gongti E. Road, Chaoyang District
Beijing 100020   Tel. (10) 6506-6688 Ext. 8402
                            (10) 6506-3101

CHINESE TAIPEI – TAIPEI CHINOIS
Good Faith Worldwide Int'l. Co. Ltd.
9th Floor, No. 118, Sec. 2
Chung Hsiao E. Road
Taipei   Tel. (02) 391.7396/391.7397
                                Fax: (02) 394.9176

**CZECH REPUBLIC –
RÉPUBLIQUE TCHÈQUE**
National Information Centre
NIS – prodejna
Konviktská 5
Praha 1 – 113 57   Tel. (02) 24.23.09.07
                                Fax: (02) 24.22.94.33
E-mail: nkposp@dec.niz.cz
Internet: http://www.nis.cz

DENMARK – DANEMARK
Munksgaard Book and Subscription Service
35, Nørre Søgade, P.O. Box 2148
DK-1016 København K   Tel. (33) 12.85.70
                                Fax: (33) 12.93.87

J. H. Schultz Information A/S,
Herstedvang 12,
DK – 2620 Albertslung   Tel. 43 63 23 00
                                Fax: 43 63 19 69
Internet: s-info@inet.uni-c.dk

EGYPT – ÉGYPTE
The Middle East Observer
41 Sherif Street
Cairo   Tel. (2) 392.6919
                                Fax: (2) 360.6804

FINLAND – FINLANDE
Akateeminen Kirjakauppa
Keskuskatu 1, P.O. Box 128
00100 Helsinki

Subscription Services/Agence d'abonnements :
P.O. Box 23
00100 Helsinki   Tel. (358) 9.121.4403
                            Fax: (358) 9.121.4450

***FRANCE**
OECD/OCDE
Mail Orders/Commandes par correspondance :
2, rue André-Pascal
75775 Paris Cedex 16   Tel. 33 (0)1.45.24.82.00
                         Fax: 33 (0)1.49.10.42.76
                            Telex: 640048 OCDE
Internet: Compte.PUBSINQ@oecd.org

Orders via Minitel, France only/
Commandes par Minitel, France exclusivement :
36 15 OCDE

OECD Bookshop/Librairie de l'OCDE :
33, rue Octave-Feuillet
75016 Paris   Tel. 33 (0)1.45.24.81.81
                             33 (0)1.45.24.81.67

Dawson
B.P. 40
91121 Palaiseau Cedex   Tel. 01.89.10.47.00
                              Fax: 01.64.54.83.26

Documentation Française
29, quai Voltaire
75007 Paris   Tel. 01.40.15.70.00

Economica
49, rue Héricart
75015 Paris   Tel. 01.45.78.12.92
                              Fax: 01.45.75.05.67

Gibert Jeune (Droit-Économie)
6, place Saint-Michel
75006 Paris   Tel. 01.43.25.91.19

Librairie du Commerce International
10, avenue d'Iéna
75016 Paris   Tel. 01.40.73.34.60

Librairie Dunod
Université Paris-Dauphine
Place du Maréchal-de-Lattre-de-Tassigny
75016 Paris   Tel. 01.44.05.40.13

Librairie Lavoisier
11, rue Lavoisier
75008 Paris   Tel. 01.42.65.39.95

Librairie des Sciences Politiques
30, rue Saint-Guillaume
75007 Paris   Tel. 01.45.48.36.02

P.U.F.
49, boulevard Saint-Michel
75005 Paris   Tel. 01.43.25.83.40

Librairie de l'Université
12a, rue Nazareth
13100 Aix-en-Provence   Tel. 04.42.26.18.08

Documentation Française
165, rue Garibaldi
69003 Lyon   Tel. 04.78.63.32.23

Librairie Decitre
29, place Bellecour
69002 Lyon   Tel. 04.72.40.54.54

Librairie Sauramps
Le Triangle
34967 Montpellier Cedex 2   Tel. 04.67.58.85.15
                              Fax: 04.67.58.27.36

A la Sorbonne Actual
23, rue de l'Hôtel-des-Postes
06000 Nice   Tel. 04.93.13.77.75
                              Fax: 04.93.80.75.69

GERMANY – ALLEMAGNE
OECD Bonn Centre
August-Bebel-Allee 6
D-53175 Bonn   Tel. (0228) 959.120
                           Fax: (0228) 959.12.17

GREECE – GRÈCE
Librairie Kauffmann
Stadiou 28
10564 Athens   Tel. (01) 32.55.321
                              Fax: (01) 32.30.320

HONG-KONG
Swindon Book Co. Ltd.
Astoria Bldg. 3F
34 Ashley Road, Tsimshatsui
Kowloon, Hong Kong   Tel. 2376.2062
                              Fax: 2376.0685

HUNGARY – HONGRIE
Euro Info Service
Margitsziget, Európa Ház
1138 Budapest   Tel. (1) 111.60.61
                              Fax: (1) 302.50.35
E-mail: euroinfo@mail.matav.hu
Internet: http://www.euroinfo.hu//index.html

ICELAND – ISLANDE
Mál og Menning
Laugavegi 18, Pósthólf 392
121 Reykjavik   Tel. (1) 552.4240
                              Fax: (1) 562.3523

INDIA – INDE
Oxford Book and Stationery Co.
Scindia House
New Delhi 110001   Tel. (11) 331.5896/5308
                              Fax: (11) 332.2639
E-mail: oxford.publ@axcess.net.in

17 Park Street
Calcutta 700016   Tel. 240832

INDONESIA – INDONÉSIE
Pdii-Lipi
P.O. Box 4298
Jakarta 12042   Tel. (21) 573.34.67
                              Fax: (21) 573.34.67

IRELAND – IRLANDE
Government Supplies Agency
Publications Section
4/5 Harcourt Road
Dublin 2   Tel. 661.31.11
                              Fax: 475.27.60

ISRAEL – ISRAËL
Praedicta
5 Shatner Street
P.O. Box 34030
Jerusalem 91430   Tel. (2) 652.84.90/1/2
                              Fax: (2) 652.84.93

R.O.Y. International
P.O. Box 13056
Tel Aviv 61130   Tel. (3) 546 1423
                              Fax: (3) 546 1442
E-mail: royil@netvision.net.il

Palestinian Authority/Middle East:
INDEX Information Services
P.O.B. 19502
Jerusalem   Tel. (2) 627.16.34
                              Fax: (2) 627.12.19

ITALY – ITALIE
Libreria Commissionaria Sansoni
Via Duca di Calabria, 1/1
50125 Firenze   Tel. (055) 64.54.15
                              Fax: (055) 64.12.57
E-mail: licosa@ftbcc.it

Via Bartolini 29
20155 Milano   Tel. (02) 36.50.83

Editrice e Libreria Herder
Piazza Montecitorio 120
00186 Roma   Tel. 679.46.28
                              Fax: 678.47.51

Libreria Hoepli
Via Hoepli 5
20121 Milano   Tel. (02) 86.54.46
                              Fax: (02) 805.28.86

Libreria Scientifica
Dott. Lucio de Biasio 'Aeiou'
Via Coronelli, 6
20146 Milano   Tel. (02) 48.95.45.52
                                Fax: (02) 48.95.45.48

JAPAN – JAPON
OECD Tokyo Centre
Landic Akasaka Building
2-3-4 Akasaka, Minato-ku
Tokyo 107   Tel. (81.3) 3586.2016
                                Fax: (81.3) 3584.7929

KOREA – CORÉE
Kyobo Book Centre Co. Ltd.
P.O. Box 1658, Kwang Hwa Moon
Seoul   Tel. 730.78.91
                                Fax: 735.00.30

MALAYSIA – MALAISIE
University of Malaya Bookshop
University of Malaya
P.O. Box 1127, Jalan Pantai Baru
59700 Kuala Lumpur
Malaysia   Tel. 756.5000/756.5425
                                Fax: 756.3246

MEXICO – MEXIQUE
OECD Mexico Centre
Edificio INFOTEC
Av. San Fernando no. 37
Col. Toriello Guerra
Tlalpan C.P. 14050
Mexico D.F.   Tel. (525) 528.10.38
                                Fax: (525) 606.13.07
E-mail: ocde@rtn.net.mx

NETHERLANDS – PAYS-BAS
SDU Uitgeverij Plantijnstraat
Externe Fondsen
Postbus 20014
2500 EA's-Gravenhage   Tel. (070) 37.89.880
Voor bestellingen: Fax: (070) 34.75.778

Subscription Agency/   Agence d'abonnements :
SWETS & ZEITLINGER BV
Heereweg 347B
P.O. Box 830
2160 SZ Lisse   Tel. 252.435.111
                                Fax: 252.415.888

NEW ZEALAND – NOUVELLE-ZÉLANDE
GPLegislation Services
P.O. Box 12418
Thorndon, Wellington   Tel. (04) 496.5655
                                Fax: (04) 496.5698

NORWAY – NORVÈGE
NIC INFO A/S
Ostensjoveien 18
P.O. Box 6512 Etterstad
0606 Oslo   Tel. (22) 97.45.00
                                Fax: (22) 97.45.45

PAKISTAN
Mirza Book Agency
65 Shahrah Quaid-E-Azam
Lahore 54000   Tel. (42) 735.36.01
                                Fax: (42) 576.37.14

PHILIPPINE – PHILIPPINES
International Booksource Center Inc.
Rm 179/920 Cityland 10 Condo Tower 2
HV dela Costa Ext cor Valero St.
Makati Metro Manila   Tel. (632) 817 9676
                                Fax: (632) 817 1741

POLAND – POLOGNE
Ars Polona
00-950 Warszawa
Krakowskie Prezdmiescie 7   Tel. (22) 264760
                                Fax: (22) 265334

PORTUGAL
Livraria Portugal
Rua do Carmo 70-74
Apart. 2681
1200 Lisboa   Tel. (01) 347.49.82/5
                                Fax: (01) 347.02.64

SINGAPORE – SINGAPOUR
Ashgate Publishing
Asia Pacific Pte. Ltd
Golden Wheel Building, 04-03
41, Kallang Pudding Road
Singapore 349316   Tel. 741.5166
                                Fax: 742.9356

SPAIN – ESPAGNE
Mundi-Prensa Libros S.A.
Castelló 37, Apartado 1223
Madrid 28001   Tel. (91) 431.33.99
                                Fax: (91) 575.39.98
E-mail: mundiprensa@tsai.es
Internet: http://www.mundiprensa.es

Mundi-Prensa Barcelona
Consell de Cent No. 391
08009 – Barcelona   Tel. (93) 488.34.92
                                Fax: (93) 487.76.59

Libreria de la Generalitat
Palau Moja
Rambla dels Estudis, 118
08002 – Barcelona
                (Suscripciones) Tel. (93) 318.80.12
                (Publicaciones) Tel. (93) 302.67.23
                                Fax: (93) 412.18.54

SRI LANKA
Centre for Policy Research
c/o Colombo Agencies Ltd.
No. 300-304, Galle Road
Colombo 3   Tel. (1) 574240, 573551-2
                                Fax: (1) 575394, 510711

SWEDEN – SUÈDE
CE Fritzes AB
S–106 47 Stockholm   Tel. (08) 690.90.90
                                Fax: (08) 20.50.21

For electronic publications only/
Publications électroniques seulement
STATISTICS SWEDEN
Informationsservice
S-115 81 Stockholm   Tel. 8 783 5066
                                Fax: 8 783 4045

Subscription Agency/Agence d'abonnements :
Wennergren-Williams Info AB
P.O. Box 1305
171 25 Solna   Tel. (08) 705.97.50
                                Fax: (08) 27.00.71

Liber distribution
Internatinal organizations
Fagerstagatan 21
S-163 52 Spanga

SWITZERLAND – SUISSE
Maditec S.A. (Books and Periodicals/Livres
et périodiques)
Chemin des Palettes 4
Case postale 266
1020 Renens VD 1   Tel. (021) 635.08.65
                                Fax: (021) 635.07.80

Librairie Payot S.A.
4, place Pépinet
CP 3212
1002 Lausanne   Tel. (021) 320.25.11
                                Fax: (021) 320.25.14

Librairie Unilivres
6, rue de Candolle
1205 Genève   Tel. (022) 320.26.23
                                Fax: (022) 329.73.18

Subscription Agency/Agence d'abonnements :
Dynapresse Marketing S.A.
38, avenue Vibert
1227 Carouge   Tel. (022) 308.08.70
                                Fax: (022) 308.07.99

See also – Voir aussi :
OECD Bonn Centre
August-Bebel-Allee 6
D-53175 Bonn (Germany)   Tel. (0228) 959.120
                                Fax: (0228) 959.12.17

THAILAND – THAÏLANDE
Suksit Siam Co. Ltd.
113, 115 Fuang Nakhon Rd.
Opp. Wat Rajbopith
Bangkok 10200   Tel. (662) 225.9531/2
                                Fax: (662) 222.5188

TRINIDAD & TOBAGO, CARIBBEAN TRINITÉ-ET-TOBAGO, CARAÏBES
Systematics Studies Limited
9 Watts Street
Curepe
Trinidad & Tobago, W.I.   Tel. (1809) 645.3475
                                Fax: (1809) 662.5654
E-mail: tobe@trinidad.net

TUNISIA – TUNISIE
Grande Librairie Spécialisée
Fendri Ali
Avenue Haffouz Imm El-Intilaka
Bloc B 1 Sfax 3000   Tel. (216-4) 296 855
                                Fax: (216-4) 298.270

TURKEY – TURQUIE
Kültür Yayinlari Is-Türk Ltd.
Atatürk Bulvari No. 191/Kat 13
06684 Kavaklidere/Ankara
                                Tel. (312) 428.11.40 Ext. 2458
                                Fax : (312) 417.24.90
Dolmabahce Cad. No. 29
Besiktas/Istanbul   Tel. (212) 260 7188

UNITED KINGDOM – ROYAUME-UNI
The Stationery Office Ltd.
Postal orders only:
P.O. Box 276, London SW8 5DT
Gen. enquiries   Tel. (171) 873 0011
                                Fax: (171) 873 8463

The Stationery Office Ltd.
Postal orders only:
49 High Holborn, London WC1V 6HB
Branches at: Belfast, Birmingham, Bristol,
Edinburgh, Manchester

UNITED STATES – ÉTATS-UNIS
OECD Washington Center
2001 L Street N.W., Suite 650
Washington, D.C. 20036-4922 Tel. (202) 785.6323
                                Fax: (202) 785.0350
Internet: washcont@oecd.org

Subscriptions to OECD periodicals may also be
placed through main subscription agencies.

Les abonnements aux publications périodiques de
l'OCDE peuvent être souscrits auprès des
principales agences d'abonnement.

Orders and inquiries from countries where Distribu-
tors have not yet been appointed should be sent to:
OECD Publications, 2, rue André-Pascal, 75775
Paris Cedex 16, France.

Les commandes provenant de pays où l'OCDE n'a
pas encore désigné de distributeur peuvent être
adressées aux Éditions de l'OCDE, 2, rue André-
Pascal, 75775 Paris Cedex 16, France.

                                               12-1996

OECD PUBLICATIONS, 2, rue André-Pascal, 75775 PARIS CEDEX 16
PRINTED IN FRANCE
(58 97 02 1 P) ISBN 92-64-15613-5 – No. 49737 1997